Joseph
Comparative physical
biology.

Comparative Physical Biology

Comparative Physical Biology

N. R. Joseph

University of Illinois at the Medical Center, Chicago, Ill.

With 7 figures and 45 tables, 1973

S. Karger · Basel · München · Paris · London · New York · Sydney

S. Karger · Basel · München · Paris · London · New York · Sydney
Arnold-Böcklin-Strasse 25, CH-4011 Basel (Switzerland)

Printed in Switzerland by City-Druck, Glattbrugg
ISBN 3-8055-1485-9

Contents

Preface

The present volume attempts to develop a conceptual scheme for Comparative Physical Biology that is to a large extent based on CARNOT's principle, GIBBS' 'equilibrium of heterogeneous substances' and CLAUDE BERNARD's concept of the functions of a constant *milieu intérieur*. In the 20th century, essentially the same ideas were the basis of L. J. HENDERSON's treatment of the functions of blood (1928).

To apply this conceptual scheme to the structure and behavior of cells and tissues in general requires the application of phase rule to all possible states of aggregation. Biological processes based on contractility and irritability can then be studied as changes of state of colloid, water and electrolytes.

These studies would not have been possible without the cooperation, support and encouragement that have been generously provided by H. R. CATCHPOLE (Department of Pathology) and M. B. ENGEL (Department of Orthodontics), Colleges of Medicine and Dentistry, University of Illinois, Chicago.

It is a pleasure to acknowledge these favors.

NORMAN R. JOSEPH,
University of Illinois at the Medical Center, Chicago, Ill.

Comparative Physical Biology

Bedaekleon: Hallo! Who's there?
Philocleon: Nothing but molecules.

(Aristophanes: 'The Wasps', free translation)

Chapter I: Fundamental Concepts

In the phenomena of life, two main types of processes may be distinguished: reversible and irreversible. These correspond respectively to the continuities and discontinuities of natural history. The two types may be described by the laws of nature as comprehended by the human mind. Biological continuity depends on the conservation of constant conditions. Constancy in the organism is sustained by reversible processes, and by conditions of thermodynamic invariance and constraint. Development and change are generally irreversible. Biological continuity may therefore be described by a principle which may be called the 'conservation of reversibility'. This principle explains the continuous existence of species over varying periods of time, extending in some instances to hundreds of millions of years.

In such species, biological stability implies not only extreme endurance of the units of heredity – genes and chromosomes – but also the endurance in time of physicochemical principles of invariance and constraint, as applied to water and the inorganic electrolytes. Biological stability over any period of time implies a limited number of degrees of freedom in the physicochemical system, borrowing the terms from thermodynamics and the phase rule [JOSEPH, 1971a, b]. This explains the endurance of complex, coherent systems which maintain permanent kinds of relationships with the cosmic environment [BERNARD, 1878; MACALLUM, 1910, 1926].

The enduring patterns of life are not absolutely constant, since the cosmic environment shows slow irreversible changes on a geological time scale. For example, the oceans change composition through gradual increases of all the biologically important inorganic salts. Mainly these consist of the chlorides of sodium, potassium, calcium and magnesium. The concentrations of the ions increase at varying rates. Evidence shows that magnesium increases more rapidly than calcium, and that sodium increases more rapidly than potassium. All salts are drained into the ocean by the continuous flow of rivers from the five continents. Sodium salts increase at a faster rate, because potassium is retained by living plants and animals and by the soil, which is rich in decomposed biological material. Magnesium increases rapidly because, unlike calcium, it forms few insoluble salts. Therefore, it has little tendency to enter the bony parts of vertebrates or the shells of invertebrates. In general, progressive changes of the salinity of the ocean have compelled its vertebrates and invertebrates to adapt themselves to the irreversible drift of the environment. This must have resulted in an irreversible shift in the population and ecology of the oceans. Since the salt composition varies greatly in different parts of the ocean, and especially in estuaries, adaptations in certain species would have required only simple migrations to favorable habitats.

Any treatment of 'comparative physical biology' must account for the stability of living species in an irreversibly changing environment and for irreversible genetic changes in a constant environment. Ontogenetic and phylogenetic development may take place under constant environmental conditions, as well as under constant internal conditions. Constant external and internal conditions tend to limit the kinds of variations which are free to develop in living cells. In mammalian ontogeny, the restrictions are imposed by the stability and invariance of the genes, and by the constancy of the *milieu intérieur.* In phylogeny, there are irreversible genetic mutations, but survival of the mutant forms depends on adaptation to a favorable environment. Phylogenetic changes are limited by the invariant nature of the inorganic electrolytes, which are 'eternal objects', and which cannot evolve.

This limitation imposes thermodynamic conditions of invariance and constraint within any organism. These conditions limit phylogenetic development to certain possible channels with respect to electrolyte and water composition. This applies to all cells and tissues, including the *milieu intérieur.* Sodium, potassium and chloride cannot react irreversibly with cells and tissues. They enter only into reversible processes. The invariance of the chemical morphology of any tissue at any period of life depends on the conservation of reversibility of the vital processes.

The physical principles which restrict and determine chemical morphology are thermodynamic, and depend on the invariant nature of water and the inorganic ions in the body fluids. In mammals and birds, blood plasma and other body fluids are confined to quite narrow ranges of water and electrolyte composition. In fish and invertebrates, the *milieu intérieur* is less constant, but it will be shown that there are nevertheless very strict limitations on its variability. In all cases, physicochemical variability is described by thermodynamic conditions of invariance and constraint, which limit the number of degrees of freedom in any system of cells and tissues.

Thermodynamic conditions operate within systems that are controlled also by strict conditions of biological constraint. In any species at any period of the life span, constraints are imposed by the nature of the genes and chromosomes which control the synthesis of well-ordered sets of intracellular and extracellular macromolecular substances. All cells and tissues grow and develop, therefore, under two sets of controls – thermodynamic and genetic. Chemical composition of the intracellular protoplasm and the extracellular and membranous connective tissue cannot be free and independent in the normally developing animal organism. It is well-ordered, constrained or channeled into the normal paths of growth and development.

Always, at any period of life, there are accessible and inaccessible states of the entire physicochemical system. Normal growth and development is confined to the most accessible states, which can develop in only one normal way. This implies that the organism is univariant, with one degree of freedom. The restriction to the most accessible or most probable state is brought about by the laws of thermodynamics and statistical mechanics. These are laws of probability rather than of mechanical determinism.

In periods of rapid morphological change, as in the growth period of mammals, development may be considered to be univariant. There is only one degree of freedom in any type of tissue, or only one mode of normal growth and development. After the end of the growth period, the organism may approach a state of invariance. The system is then characterized as having zero degrees of freedom. In this period, cells and tissues would be characterized by constant chemical composition. They could also be characterized by constant physicochemical, physiological and metabolic properties, all of which depend on chemical morphology and thermodynamic state. State is determined by the embryological and ontogenetic development of well-ordered sets of proteins, carbohydrates and lipids of each kind of tissue. This occurs under conditions imposed by the constant electrolyte composition of the *milieu intérieur.* This consists mainly of six substances: water and ionic sodium, potassium, calcium, magnesium and chloride.

Normal channeling of growth, development and aging depends on the reversibility of all processes into which these substances can enter. Water and the five physiological ions are normally found in practically all cells and tissues of the body. Then any cell or tissue requires for its description the composition with respect to (n + 6) substances. Here *n* denotes the number of characteristic intercellular or intracellular macromolecular substances, and 6 denotes water and the five inorganic ions. This is a minimal number of substances that are required to describe the chemical composition of any phase.

Physicochemical state requires additional variables for its description. All variables depend on chemical morphology and state, and are therefore univariant or invariant in time. At any period of ontogenetic development, there is only one normal state that characterizes the system. This may be denoted as the standard resting state, the state of maximal stability, or the most accessible state. However, there are usually other states, which may be denoted as normally accessible states. These represent states of stimulation, modified metabolism, respiration or physiological activity of various kinds.

Depending on the nature of the cell or tissue, the existence of these accessible states represents an increase in the number of degrees of freedom. Except for tissues such as heart muscle, which contracts periodically and which is never at rest, most structures remain for appreciable time in the resting standard state, and are characterized by zero degrees of freedom. Heart muscle, on the other hand, changes from diastole to systole with each beat. If each limiting state is normal and constant, heart is characterized by one degree of freedom. Circulating blood, which is kept in a state of continuous flow by the heart, is also characterized by one degree of freedom: it also exists in two limiting states, alternating cyclically between arterial and venous blood. Physiological processes are therefore optimally regular, well-ordered, and characterized by a small number of degrees of freedom. Changes of state may be brought about either by changes of the *milieu intérieur,* or by changes in chemical morphology and state of the (n + 6) intracellular or extracellular substances.

Aims and Methods

Without clear definitions of the fundamental aims of the study, it would seem to be almost impossible to produce a rational correlation of the physical, chemical and physiological properties of cells and tissues. To relate these properties to phylogenetic development and speciation would also be impossible. The first requirement for logical development of the subject would seem to be the development of strict criteria for a logical conceptual scheme. This involves clear distinctions in all concepts that involve the subject-predicate relationship. Without such clear distinctions, logical thought becomes impossible. Inductive generalizations and hypothetical formulations require at every step a clear statement of the subject-predicate relationship; innumerable serious errors in biology have resulted from failure to observe those necessary criteria. Non-logical or illogical thought can yield only erroneous and non-logical conclusions. This would seem to be self-evident, but a survey of recent physiological literature shows that much of the reasoning is non-logical and self-contradictory.

In examining the general principles of subject-predicate relationships, we assume that clear and distinct ideas can always be expressed by simple declarative sentences, and that affirmative sentences lead to the most useful deductive generalizations. Negative premises are useful and important, but of themselves do not lead to positive or affirmative conclusions. The most important positive statements are those that begin with the subject.

If, for example, we take the man Shakespeare as subject, any scholar familiar with English drama can immediately produce an indefinitely large number of predicates, both positive and negative. 'He was a great English poet and dramatist of the Elizabethan period.' One subject immediately suggests five predicates – great, English, poet, dramatist and Elizabethan. The five predicates suggest 'Shakespeare' to any schoolboy, but the omission of any one attribute would greatly widen the range of poets who could qualify as possible subjects. In any case, the identification of Shakespeare is somewhat uncertain without further information. It is an example of 'hypothesis' or 'synthetic inference' rather than of deduction or 'analytical inference'.

In setting up any conceptual scheme as a foundation for physical biology, the definition of the subject matter is a most important step. Without a clear distinction between subject and predicate in every step of a chain of logical thought, cause and effect almost inevitably become confused, and logical fallacies are produced in great profusion. Induction, another form of synthetic inference, could lead to the conclusion that all great English plays of the Elizabethan period were written by Shakespeare. This conclusion may be true, but it is illogical, since it does not follow from the premises.

In the following chapters, it will be necessary to point out many examples of non-logical reasoning that have been used in the development of current concepts in physiology and biology. These tend to accumulate at an exponential rate, leading to no well-ordered development, but rather to an impressive array of non-verifiable 'facts' and conclusions. In a memorable passage, WHITEHEAD [1925] has described such accumulations as 'scrapheaps'.

To be useful in science, hypotheses must be verifiable. In the biological sciences, many non-verifiable hypotheses have been retained as foundation material or as subject matter, even when it has been shown that the reasoning contradicts established physical principles. In most cases, the contradictions result from non-logical confusion of subject and predicate, and an abuse of synthetic inference. Sound reasoning, on the other hand, must be based on a clear choice of subject in every case, and strict inferences regarding the relationship between primary functions and secondary properties.

Confusion between cell morphology and the secondary properties of the cell always leads to an inversion of the cause-effect relationship, and to an inversion of subject and predicate. Confusion and contradiction can be avoided by the realization of the simple fact that the primary function of any cell or tissue is to exist. This is the necessary condition for it to have any other function. Existence is determined by the embryological and ontogenetic development of cell morphology and structure. This determines a set of primary properties on which all physicochemical and physiological properties must depend [JOSEPH, 1971a, b]. It is fallacious to attempt to derive the chemical morphology of the cell from its secondary properties. To do this is to invert the subject-predicate relationship and to confuse causes and effects.

To obtain valid derivations of the secondary properties of cells from chemical morphology, it is necessary to define the minimal elementary properties that are required. Chemical morphology is described by the composition of the protoplasmic constituents and by the content of water and electrolytes. Determination of total water and electrolytes may be expressed as g/kg of fresh tissue or cellular material. The remaining mass could be expressed as colloid or protoplasmic mass per kg of tissue. When the ion concentrations are expressed in mol/kg water: $(c_{Na})''$, $(c_K)''$, $(c_{Ca})''$, $(c_{Mg})''$ and $(c_{Cl})''$, there are $(n + 6)$ quantities that are necessary to express the chemical composition. Here n denotes the number of macromolecular substances, and 6 represents water and the five species of inorganic ions. However, a detailed description of the n intracellular colloidal substances is almost never available. For practical purposes, the analysis of water and the five ions suffices to give a useful description of the chemical morphology. Colloid may then be described as g/kg water. Its physicochemical properties are then expressed as average values for the entire colloidal mass, i.e. as mean equivalent weight of the total colloid.

In thermodynamics, physicochemical state is expressed in terms of chemical composition, temperature and pressure. Thus the state of a biological structure would be expressed in relation to (n +8) quantities, where the composition is expressed by (n + 6) quantities. The other two parameters refer to temperature and pressure. The set of (n + 8) primary quantities describe cellular morphology and state, which are generally invariant or univariant in time. All the secondary properties constitute a set of attributes, each member of which is a function of the (n + 8) members of the primary set. If the primary set is denoted as A, B, C . . ., and the secondary set as X, Y, Z . . ., then:

$$A, B, C \ldots \supset X, Y, Z \ldots \tag{1.1}$$

The ⊃ sign may be read as 'implies'. Thus the symbols signify that 'A, B, C . . .' implies 'X, Y, Z . . .', or that 'X, Y, Z . . .' depends on 'A, B, C . . .'. Here A, B, C . . . refer to temperature, pressure and composition of the system, and X, Y, Z . . . refer to an indefinitely large number of secondary properties. These would include the resting electrical potential or electromotive force, the action potential, the respiratory rate, expressed as O_2 consumption, CO_2 production, or heat production in calories, permeabilities, muscular work and tension, and so forth.

The resting rate of metabolism is treated as a secondary rather than as a primary property. In cold-blooded animals it depends on the absolute temperature T:

$$T \supset M \tag{1.2}$$

Fish and invertebrates always come to the temperature of the environment. Thus it is incorrect to write

$$M \supset T \tag{1.3}$$

which would imply that the temperature is controlled by the metabolic rate. In laboratory experimentation, preparations of cells and tissues are placed in the cold room at 0 °C. to produce a low rate of metabolism. The relation 1.3 would imply that the temperature of the cold room was determined by the metabolism of the cells.

In the literature, such experiments have been performed to prove 'active transport' or the existence of a metabolically dependent 'sodium pump'. The conclusion could be expressed in the form:

$$A, B, C, T \subset M \tag{1.4}$$

This would mean that the electrolyte composition and temperature of the cell are determined by metabolic processes that transport electrolytes.

Actually, the logical relationships are expressed by

$$A, B, C, T \supset M \tag{1.5}$$

and

$$A, B, C, M \subset T \tag{1.6}$$

These imply the correct relation between M and T, as given by 1.2. To arrive at the false conclusion 1.4 requires the omission of T as the independent variable. In formal logic, the fallacy is that of 'omitting the middle term'. In the syllogism, T is replaced by M, and the physicochemical changes appear to be governed by changes in M. The real independent variable, T, is thus replaced by the dependent variable, M, and all the subsequent reasoning is falsified. This is an example of the way in which subject-predicate and cause-effect relationships have been inverted in the theory of 'active transport'.

Two fundamental aims and methods of the subject thus need to be revised and clarified. First, all logical relationships between primary and secondary properties of cells and tissues require to be defined. Secondly, the kinds of valid analytical and experimental data to be admitted must be critically formulated from the point of view of physical chemistry, thermodynamics and general biology.

Distribution and Transport

It is necessary to distinguish the *distribution* of electrolytes from their *transport,* and to decide which is primary and which is secondary. By 'distribution' is meant the most stable composition in a heterogeneous biological system. Change of state to a secondary state of normal accessibility, as in stimulation or change of metabolism, would lead to a *re-distribution* of ions. This is due to a change of chemical morphology of a configurational nature in one or more phases. Thus distribution of ions and water is *identified* with morphological state, rather than *caused* by state. It is therefore a primary property of the cell or tissue. It is not *caused* by the resting metabolism, permeability, or any other secondary function of the system, but is implied by the existence of the cell as a morphological or anatomical unit.

Transport, on the other hand, is a secondary rather than a primary function. In the resting standard state of maximal stability, transport may be *reversible* or *irreversible,* depending on the nature of the electrolyte. Irreversible transport of sodium, potassium or chloride would result in a change of composition of the phases that are involved; it is thus inconsonant with the conditions necessary for an invariant standard state. On the other hand, irreversible transport of carbonic acid is a necessary condition for the maintenance of an invariant steady state in actively metabolizing cells.

Homeostasis, or physiological invariance, requires that carbonic acid be transported irreversibly to circulating blood or hemolymph at the same rate at which it is formed. It is then irreversibly transported to the external environment by circulation and expiration. In warm blooded species, heat production occurs at the same rate as respiration, measured in calories. Irreversible transport of heat to the environment then occurs at the same rate as heat production or respiration. The necessary conditions for an invariant steady state within the body thus require constant rates of respiratory metabolism and irreversible transport of metabolites, nutrients and heat. All transport rates would be coordinated with the rates of cellular respiration and heat production. Invariant composition of arterial and venous blood then imply constant rates of circulation and pulmonary ventilation.

Rates of irreversible transport are necessary functions which are determined by the standard distributions of electrolytes and water in the various cells and tissues. The standard distributions are seen to be synonymous with the standard resting chemical morphology of the tissues. Normal standard distributions of water and electrolytes are sustained by two kinds of transport. The first class includes ions of environmental origin – sodium, potassium calcium, magnesium and chloride. These cannot be produced or consumed by cells and tissues. They are normal constituents of sea water, and of all biological fluids necessary for the normal growth and maintenance of vertebrate cells and tissues. According to MACALLUM [1910, 1926], they are to be traced back to earlier states of sea water, or to the primitive waters in which life may be thought to have originated. On the geological time scale, natural biological fluids may be considered to have preceded the existence of living forms [JOSEPH, 1971b].

The second class of electrolytes includes those which enter cellular metabolism, either as reactants or as products. These would include, for example, amino acids or other nitrogenous substances, and carbon dioxide or carbonic acid, as produced in aerobic respiration. Lactic acid, produced in anaerobic respiration, is also an example of the second class of electrolytes. The condition of invariance or homeostasis would require the irreversible transport of electrolytes of the second class.

Ions of the first class, including sodium, potassium and chloride, are transported reversibly. In the standard state, the rates of transport are zero. It is thus superfluous to speak of transport. The work of ion transport and the efficiency of ion transport mechanisms are also zero. This follows from the second law of thermodynamics (CARNOT's principle), according to which isothermal reversible work is zero in an invariant system [JOSEPH, 1966a]. Applying the principle known as Occam's razor, ion pumps and carriers should be excluded from the conceptual scheme ('entities are not to be multiplied without necessity').

The criterion for a reversible process in an invariant system is:

$$\Delta G = 0 \tag{1.7}$$

where ΔG is the *free energy change*. This corresponds to the maximal work of any reversible process, which is also zero. The idea of a mechanism for such processes is unnecessary, and according to Occam's razor, should not be introduced.

For irreversible processes, the criterion is:

$$\Delta G < 0 \tag{1.8}$$

In the combustion of 1 mol of glucose:

$$C_6H_{12}O_6 + 6\,O_2 = 6\,CO_2 + 6\,H_2O$$

the value of ΔG is $-690{,}500$ cal under physiological conditions. According to 1.8, the formation of 6 molecules of H_2CO_3, attended by the liberation of 690,500 cal, is a spontaneous reaction, which requires no other source of free energy.

Transport of extracellular oxygen and glucose is part of the same process. All transport processes are therefore spontaneous and proceed with the diminution of free energy that occurs in the course of the reaction. Transport mechanisms are therefore superfluous in the conceptual scheme and should be excluded.

According to 1.7 and 1.8, all transport processes are either reversible or irreversible. The rates of reversible processes are zero in the invariant standard state. Then the concept of transport and transport mechanisms becomes superfluous and unnecessary. This is true also of permeability, which is related to transport. Rates of irreversible transport are determined by the metabolic rates for reactions such as the oxidation of glucose.

Provided that these rates do not exceed the upper limits for transport rates, the condition for irreversible transport 1.8 is sufficient to maintain an invariant state of homeostasis within the cell. Here again, it is unnecessary to include transport mechanisms or membrane permeabilities in the conceptual scheme, as it applies to the standard resting state.

Since permeability, irreversible transport and respiratory metabolism are secondary functions of morphology and state, all changes of primary and secondary functions must be coordinated with each other. Thus permeability, transport and metabolism cannot vary independently as members of the primary set A, B, C... In a response which involves only a change to a normally accessible state defined as A′, B′, C′, the secondary properties would be defined as X′, Y′, Z′... Such a system is characterized by one degree of freedom. Normal states of systole and diastole in heart muscle could serve as an example.

Protoplasm

'Let us now pass to the consideration of that special substance, or rather class of substances, whose properties form the chief subject of botany and of zoology, as truly as those of the silicates form the chief subject of mineralogy: I mean the life slimes or protoplasm. Let us begin by cataloguing the general characters of the slimes. They one and all exist in two states of aggregation, a solid or nearly solid state and a liquid or nearly liquid state; but they do not pass from the former to the latter by ordinary fusion... Protoplasm, when quiescent, is, broadly speaking, solid; but when it is disturbed in an appropriate way, or sometimes even spontaneously without external disturbance, it becomes, broadly speaking, liquid' [PEIRCE, 1923].

This description of protoplasm or the life slimes may be compared with much earlier conjectures regarding the nature of intramuscular fluids that were written by STENSEN [1662]. 'Concerning the fluid of muscles how uncertain or rather how wholly wanting is our knowledge. Fluid certainly exists in the fibrillae of which the motor fibers are composed and between the fibrillae, also between the motor fibres themselves, in the membranous fibrillae' (that is the connective tissue), 'and between the membranous fibrillae; but in truth it is by no means clear whether these fluids are all of one kind or whether just as they are distinct in the seats which they occupy so they may differ in material properties...

'As the substance of this fluid is unknown to us, so is its movement undetermined, since neither by sure reasoning nor by experiment has it been ascertained whence it comes, whither it tends, where on its departure it betakes itself . . .

'There remains another difficulty of no less moment not yet cleared up, namely in what respect the movement of the fluid in a muscle while it is contracting differs from the movement of the fluid in the same muscle when it remains quiet, uncontracted. Is the quantity changed or does it remain the same? Is the fluid after the event, supposing it constant, the same as before the event? Does the fluid move because the muscle contracts, or does the contraction of the solid proceed from the movement of the fluid?'

Thus more than two centuries before PEIRCE's description of the liquefaction of protoplasm, STENSEN was clearly aware of the problem and stated the main implications. These are clearly related to the general problem of cellular irritability as stated first by GLISSON [1654] in the 17th century and by VON HALLER [1757] in the following century. In the two centuries between GLISSON and STENSEN, on one hand, and PEIRCE on the other, physiologists and other biologists were aware of a problem of changes of state of protoplasm in vital processes. Formulation of the problem on a physicochemical basis required the development of the thermodynamics of heterogeneous systems. On the basis of early formulations, it would seem that a two-state theory or multi-state theory of protoplasm would have been required. Yet the actual development of the subject took place along different lines.

Late in the 19th century BÜTSCHLI [1892] attempted to formulate a theory of protoplasm based on the idea of a microscopic or submicroscopic 'foam-like' structure. On histochemical grounds his views were criticized by HARDY [1899], who considered BÜTSCHLI's preparations to represent the nature of his staining methods rather than the actual intracellular structure. HARDY [1905] went on to study *in vitro* solutions of water-soluble proteins: his work is now regarded as one of the foundations of the physical chemistry of protein solutions. It is based on the theory of electrolytic dissociation and on the equilibrium of proteins in solutions that contain acids, bases and salts [COHN, 1925].

Subsequent studies of protein solutions were based on the work of DONNAN [1911], SORENSEN [1915] and LOEB [1924]. HENDERSON's [1928] studies of blood as a physicochemical system owed much to these predecessors, but also depended on general theoretical considerations derived from BERNARD [1878] and GIBBS [1928]. DONNAN's theoretical work is often referred to as the 'Gibbs-Donnan equilibrium'. This is quite misleading, as the 'Donnan equilibrium' is actually only a limiting case of GIBBS' more general 'equilibrium of heterogeneous substances'. DONNAN's theory applies to protein solutions which are homogeneous, and which therefore show the property of perfect mixing. It describes two-phase systems in which water exists in only one state of aggregation: the liquid state. GIBBS' more general theory applies to systems which are heterogeneous in structure, or to systems in which water exists in two or more states of aggregation. It would be applicable to muscle fibers, as described by STENSEN, or to liquefaction of protoplasm, as described by PEIRCE. Only a bi- or multi-state theory of intracellular water would be consonant with changes of irritability, as conceived by GLISSON or by VON HALLER.

Present-day theories of intracellular water have, to a great extent, been based on the one-state conception, as traditionally formulated in the DONNAN equilibrium. Such theories were criticized by FISCHER [1933] on the grounds that intracellular fluids and connective tissue fluid are not to be regarded as dilute salt solutions such as those treated in the classical theory of 'membrane equilibrium'. FISCHER regarded such systems as heterogeneous colloidal aggregates in which two or more states of water would coexist in two or more phases. This agrees with the view now accepted by many biophysicists on the basis of theoretical and experimental studies [BERENDSEN, 1968; BERNAL, 1965; KAVENAU, 1964; KLOTZ, 1965; HAZLEWOOD, 1969].

On the basis of studies of heterogeneous equilibrium, it is now possible to regard intracellular and extracellular water as existing in labile states of aggregation, determined by chemical morphology in the most accessible and normally accessible states. Physiological responses then depend on changes of state of the primary properties A, B, C . . . to a stimulated state characterized by A′, B′, C′ . . . The change of state of water would be measured by a change of one of the primary properties, for example, A to A′. The entire set of properties also changes from A, B, C . . . to A′, B′, C′ . . . The change of state would be well-ordered, and would normally occur with one degree of freedom. Then, as explained in the foregoing section, the secondary properties would change simultaneously from X, Y, Z . . . to X′, Y′, Z′ . . . The primary properties are described by the sets of $(n + 6)$ or $(n + 8)$ parameters of state. The secondary properties include changes of electrical potentials, respiratory metabolism, membrane permeability and transport, muscular work, nerve conduction, and so forth, as determined by the nature of the tissue and the nature of the stimulus.

Any such response would require a change of state of colloid and water. It would be explained only on the basis of a bi- or multi-state theory of water structure, rather than on a uni-state theory. Irritability would be incompatible with the existence of water limited to an intracellular liquid state or solution, characterized by homogeneity and perfect mixing [JOSEPH, 1971a, b]. This would limit the states of electrolytes to one standard normal state, and changes of electrical potential could not occur in a stimulated or activated state. Irritability may be thought to depend on a change of state of colloid, water and electrolytes, with a simultaneous release of configurational free energy.

Configurational Entropy

According to SCHRÖDINGER [1944], the standard resting state of living organisms is that of a highly organized heterogeneous system, in which the solid state of aggregation is maintained by nutritive supplies of 'negative entropy'. In behavioral responses, there is typically a change of the primary properties A, B, C . . . to a normally accessible state represented by A′, B′ C′ . . . The change of state is generally characterized by an increase of configurational entropy and a decrease of configurational free energy [JOSEPH, 1971a]. It is also characterized typically by a transition from a solid-like state of aggregation to a liquid or nearly liquid state [PEIRCE, 1923]. The change of state from solid to liquid is attended by a change of the dielectric properties of water as the dispersion medium for electrolytes. This gives rise to a change of electrical potential and, as in muscle, to production of work and changes of tension.

In thermodynamics, free energy, G, is related to entropy, S, by the formula:

$$G = H - TS \tag{1.9}$$

where H is the enthalpy and T is the absolute temperature. The operational quantities are then ΔG, ΔH and TΔS, which represent *changes* of G, H and TS at constant temperature. In an *ideal* system in which configurational free energy, enthalpy and entropy are zero:

$$\Delta\bar{G} = \Delta\bar{H} - T\Delta\bar{S} \tag{1.10}$$

In a real system:

$$\Delta G = \Delta H - T\Delta S \tag{1.11}$$

and

$$\Delta G_c = \Delta H_c - T\Delta S_c \tag{1.12}$$

where G_c, H_c, and S_c represent the configurational functions. In an athermal system in which ΔH is zero:

$$\Delta G = - T\Delta S \tag{1.13}$$

and

$$\Delta G_c = - T\Delta S_c \tag{1.14}$$

If, as in a rapid muscular contraction or twitch, both ΔH and ΔS approach zero, equations 1.13 and 1.14 are equivalent. This condition is approached in rapid contraction under conditions of constant composition (with no loss or gain of water, electrolytes or colloid).

Therefore, in the rapid contraction of skeletal muscle, there is a sudden decrease of configurational free energy of the fibers, leading to production of external work. In this response:

$$\Delta G = - W_{max} \tag{1.15}$$

$$\Delta G = \Delta G_c \tag{1.16}$$

and

$$\Delta G_c = - T\Delta S_c \tag{1.17}$$

Since the maximal work, W_{max}, is positive, the change of configurational entropy, ΔS_c, is also positive: ΔG and ΔG_c are negative. The resting state of the muscle is therfore characterized by low configurational entropy and high configurational free energy. In this state, the fibers are in a lengthened state. The intra- and interfibrillar phases are well-ordered, approaching the solid state of minimal entropy and maximal configurational free energy. Referred to the contracted state of high disorder and high entropy, the resting state may be described as one of negative entropy.

Intracellular states of 'negative entropy' are therefore states of high 'irritability' or 'responsiveness'. Thus STENSEN's concept of varied organization of intracellular water is easily related to GLISSON's concept of irritability. The common factor is the potential existence of the water in two or more states of entropy or free energy. According to thermodynamics, the behavioral response is that of an increase of configurational entropy and a decrease of free energy. The free energy of stretched muscle fibers is then converted to external work when muscle passes from a state of negative entropy, as in the lengthened state, to one of high entropy, as represented by the contracted state. The process is accompanied by changes of state of water and electrolytes, and by changes in all secondary properties, X, Y, Z . . .

Maintenance of a state of high irritability, *tonus*, and negative entropy requires intracellular and extracellular nutrients. Glycogen, glucose and other carbohydrates are oxidized with increases of entropy. They are therefore sources of negative entropy, and are essential in maintaining cellular irritability and tone. Fatty acids or their esters are substances of high entropy, and are oxidized with decreases of entropy. In resting metabolism, it is a condition for an invariant state of morphology that the total change of entropy must approach zero [JOSEPH, 1971a]. Thus:

$$T\Delta S = 0 \tag{1.18}$$

This requires a high coordination of all rates of combustion as a condition for homeostasis and constant states of irritability. Accordingly, the normal physiological standard state is maintained by a well-ordered input of nutrients and well-ordered respiratory metabolism. These are accompanied by highly coordinated spontaneous transport of oxygen, carbon dioxide and nutritive sources of negative and positive entropy. These are necessary to satisfy equation 1.18, which is a necessary condition for homeostasis and physiological steady states of negative configurational entropy.

Negative Entropy and Biological Irreversibility

In the animal organism, production of CO_2, H_2CO_3 and heat are spontaneous irreversible processes in which entropy increases and in which ΔH and ΔG are negative. Minimal morphological changes of cells and tissues require a null value of ΔS on a day to day basis (equation 1.19). According to equation 1.12, this requires that:

$$\Delta G = \Delta H \tag{1.19}$$

The driving force for any combustion reaction is expressed as the free energy change, ΔG; all the free energy is converted to heat. It is a condition of constant temperature in the body that all the heat is transferred to the external environment. Then $T\Delta S$ is zero, and no residual free energy of the combustion reactions is available to bring about internal chemical or physical change in the body itself. Therefore, on a day to day basis, the physicochemical state of the body remains invariant.

For normal adult man, the basal metabolic rate ($-\Delta H$) is of the order of 1,500 kcal/day [SHOCK and YIENGST, 1955]. Basal free energy ($-\Delta G$) differs from this by a factor of the order of 0–0.5%, or by less than 7 kcal/day. The conversion of calories to work units is given by:

$$1 \text{ kcal} = 41 \text{ l atm}$$

$$1 \text{ cal} = 4.183 \text{ J}$$

$$1 \text{ cal} = 0.427 \text{ kg m}$$

Expressed as internal work per day, 41 l atmospheres is of too high an order of magnitude to be credible in biological systems. Nevertheless, recent estimates on the basis of 'active transport' have led to figures of the order of 10–30% of the metabolic energy as the requirement to maintain normal intracellular electrolyte balance [GIESE, 1968]. 30% of the basal metabolic rate amounts to 450 kcal/day; the above conversion factors yield a figure of about 180,000 kgm/day to maintain the normal electrolyte concentrations. This energy is sufficient to lift a man weighing 100 kg to a height of 1.8 km (1.2 miles). The true value, calculated from the rates of reversible ion transport, is zero. This is the requirement for an invariant steady state of electrolyte balance [JOSEPH, 1966a].

Negative entropy in all cells and tissues requires nutritive sources; these originate from the photosynthesis of glucose and other carbohydrates in the green leaves of plants [GAMOW, 1965; JOSEPH, 1971a]. At the surface of the sun, photons or light quanta of high energy are emitted. The energy, $h\upsilon$, is determined by the frequency, υ, of the radiation.

$$\text{Energy} = h\upsilon \tag{1.20}$$

where h is Planck's constant.

Quantitative absorption of n light quanta by a non-reflecting 'black body' at a temperature of 300 °K (27 °C) would lead to an increase of entropy, $\Delta S_{h\upsilon}$:

$$\Delta S_{h\upsilon} = nh\upsilon\left(\frac{1}{300} - \frac{1}{6000}\right) \tag{1.21}$$

where the temperature of the sun is taken as 6,000 °K.

Thus normal solar radiation and black body absorption at the temperature of the Earth leads to an increase of entropy that is proportional to the intensity and frequency of the radiation. In the green leaves of plants, the radiant energy, $nh\upsilon$, is not absorbed as heat energy. It is converted to photochemical energy by the action of chlorophyll. This permits the chloroplasts to synthesize glucose and other high energy carbohydrates from carbon dioxide and water. Then

$$nh\nu + 6\,CO_2 + 6\,H_2O = C_6H_{12}O_6 + 6\,O_2$$

According to equation 1.22, black body absorption of light quanta leads to an immediate increase of entropy, $\Delta S_{h\nu}$. In the green leaves of plants, the synthesis of glucose from CO_2 and water is characterized by a *decrease* of entropy, in which $T\Delta S$ is about $-18{,}500$ cal. When glucose or other carbohydrates are converted to intracellular glucose or glycogen in animal nutrition, this 'negative entropy' is available for spontaneous reactions within the animal organism. In the adult mammalian organism or in man, the nutritive supplies of negative entropy are converted to heat and free energy: the conversion meets the daily energy requirements as measured by $(-\Delta H)$ and $(-\Delta G)$. Thus the conversion of nh units of photochemical energy to heat is delayed in the animal kingdom. It is not eliminated. Irreversible transport of the heat to the environment at 300 °K results in the same increase of entropy, $\Delta S_{h\nu}$, that occurs in black body absorption.

In the growth and development of the animal body, the 'negative entropy' is absorbed, and is a necessary source of muscular work, muscular 'tone' and cellular irritability. It is the source of spontaneous biological processes. After the death of the organism, it is converted by spontaneous reactions to heat and 'positive entropy'. Thanks to its vast population of living plants and animals, the planet Earth is a great pool of 'negative entropy'. In the form of biological macromolecules and ordered water, this is retained on the planet and withheld from outer space. Lifeless planets reflect most of the light quanta that they receive from the sun; this energy is transmitted to outer space, where it is ultimately absorbed by bodies that approach 0 °K. Dead planets are incapable of storing free energy as negative entropy.

Planet Earth, as long as it sustains life, can *delay* the transmission of solar energy to outer space. It thus behaves as a reservoir of negative entropy. Thus it applies a 'brake' to the operation of the second law of thermodynamics within our solar system. The fundamental irreversible processes that govern the transformation of matter and energy in the universe are modified and delayed by living organisms that inhabit the biosphere of the Earth.

In organisms of great longevity and in species of great stability, the rates of irreversible processes may approach zero. The principle of 'conservation on reversibility' is most clearly valid when applied to species in which the tempo of growth, evolution and other irreversible processes approaches zero.

In the broadest sense, the subject matter of comparative physical biology must include the edaphic and biotic requirements of all animals and plants, that is to say their complex ecological relationships. This topic is embraced by the term 'struggle for existence'. As BOLTZMANN pointed out in 1905, 'The general struggle for existence of all living beings is not the struggle for the fundamental substances, for these fundamental substances indispensable for all living creatures exist abundantly in the air, the water and the soil. The struggle is not a struggle for the energy which in the form of heat, unfortunately not available, is present in a great quantity in every object, but it is a struggle for entropy, which is available when energy passes from the hot sun to the cold earth. In order to utilize in the best manner this passage, the plants spread under the rays of the sun the immense surface of their leaves, and cause the solar energy before reaching the temperature level of the earth to make syntheses of which as yet we have no idea in our laboratories. The products of the chemical kitchen are the object of the struggle in the animal world.'

All of this becomes compatible with what has been said in the foregoing paragraphs when for BOLTZMANN's use of the term 'entropy', we substitute SCHRÖDINGER's concept of 'negative entropy'. Then in the broadest sense, man's life on Earth, as it is related to that of all other species in the biosphere, can be described as a struggle for negative entropy. This becomes, then, not only a biological, sociological and political principle, but also an historical one, as ADAMS [1919] attempted to show in applying thermodynamics to human development. Based on the authority of such thinkers as BOLTZMANN [1905], SCHRÖDINGER [1944] and LOTKA [1924], the first and second laws of thermodynamics provide a solid foundation for the principles of physical biology as well as for the science of ecology.

In attempting to formulate a unified conceptual scheme on thermodynamic rather than on the current mechanistic ideas, physical biology is faithful to the principles of general physiology as conceived by BERNARD [1878]. At the same time, it avoids, any violation of the principles of thermodynamics and heat energy, as conceived by BOLTZMANN [1905], GIBBS [1875], HELMHOLTZ [1847], and other great physicists of the 19th century. By nature, concepts of mechanistic causality are in fundamental contradiction with the laws of thermodynamics, which must be given priority over mechanism in the conceptual scheme. Mechanistic concepts, although of limited value in dealing with isolated problems on a micro scale, are not at all applicable to the physicochemical behavior of macroassemblies of molecules and systems of molecules. These can be described only by the principles of statistical mechanics, thermodynamics and the laws of probability, which transcend the physics and chemistry of isolated units of the entire physicochemical system. The behavior of the organism as a whole obeys what CANNON [1929] has termed 'the wisdom of the body' rather than the operations of what HILL [1944] has designated as 'living machinery'.

Chapter II: The Internal Environment

Within many tissues of the body the cells are surrounded by fluids and by substances in solid or semi-solid states of aggregation. Cells are in contact with other cells, with amorphous or fibrous colloidal structures and with water in one or more states of aggregation. Body water is always found to contain the mineral salts which are found in blood plasma. The cells exchange nutrients, metabolites, oxygen and carbon dioxide with the capillary circulation, which is part of the internal environment of the organism. BERNARD's term for the internal environment is *milieu intérieur*. His definition and description are classical [1878].

'The various ways in which living organisms are related to their cosmic environment enable us to study life in three forms according as existence is entirely dependent upon external conditions, less dependent, or relatively independent. These three forms of life are as follows: (1) *vie latente,* where life is not evident; (2) *vie oscillante,* where evidences of life are variable and dependent upon external environment *(milieu extérieur),* and (3) *vie constante,* where life manifests itself independently of the external environment . . . III. The third form of existence characterized by freedom and independence is found in the more highly organized animals. Here life is never suspended, but flows steadily on, apparently indifferent to alterations in its cosmic envi-ronment or changes in its material surroundings. Organs, structural mechanisms and tissues all function uniformly and their operations show no sign of the considerable variations present in organisms where conditions are inconstant. This is due to the fact that the *milieu intérieur* surrounding the organs, the tissues and their elements never varies; atmospheric changes cannot penetrate beyond it and it is therefore true to say that the physical conditions of environment are unchanging in a higher animal: each one is surrounded by this invariable *milieu,* which is, as it were, an atmosphere proper to itself in an everchanging cosmic environment. Here we have an organism which has enclosed itself in a kind of hot-house. The perpetual changes of external conditions cannot reach it; it is not subject to them, but is free and independent.

'I think I was the first to urge the belief that animals have really two environments: a *milieu extérieur* in which the organism is situated, and a *milieu intérieur* in which the tissue elements live. The living organism does not really exist in the *milieu extérieur* (the atmosphere if it breathes, salt or fresh water if that is its element) but in the liquid *milieu intérieur* formed by the circulating organic liquid which surrounds and bathes all the tissue elements; this is the lymph or plasma, the liquid part of the blood which in the higher animals, is diffused through the tissues and forms the ensemble of the intercellular liquids and is the basis of all local nutrition and the common factor of all elementary exchanges. A complex organism should be looked upon as an assemblage of simple organisms which are anatomical elements, that live in the liquid *milieu intérieur*.

'The stability of the *milieu intérieur* is the primary condition for freedom and independence of existence; the mechanism which allows of this is that which ensures in the *milieu intérieur* the maintenance of all the conditions necessary to the life of the elements. From this we know that there can be no freedom or independence of existence for simple organisms whose constituent parts are in direct contact with their cosmic environment, and that this form of life is, in fact, the exclusive possession of organisms which have attained the highest state of complexity or of organic differentiation.'

Electrolytes of Blood

The blood plasma of mammals contains the following inorganic ions: sodium, potassium, calcium, magnesium, chloride, bicarbonate, sulphate and phosphate. There is a striking similarity of composition when a comparison is made to include a variety of species. This has long been recognized. Artificial physiological saline solutions are used in the laboratory and in clinical medicine. The properties of these solutions have been experimentally investigated with respect to their effects on isolated cells and tissues. Changes in the electrolyte composition may have strongly toxic effects. A proper balance of the cations must be maintained to avoid toxicity; this necessity may be readily understood on the basis of BERNARD's principle of the constancy of the *milieu intérieur*. Body fluids or artificial substitutes must contain the required electrolytes at physiological concentrations.

The use of physiological saline solutions was investigated by RINGER [1880–1882]. The usefulness of such preparations depends on the fact that there is little variability among mammalian species when blood plasmas are compared. The electrolyte composition is nearly independent of age and of the nature of the species. The *milieu intérieur* thus tends to be an ontogenetic and a phylogenetic invariant. A 'Ringer solution' adapted to the needs of the frog is also familiar to phzsiologists. The concentrations of sodium and potassium are considerably lower than in mammalian Ringer solution, and the ratio of sodium to potassium is somewhat higher.

The blood plasma of invertebrates usually differs greatly from that of mammals, especially with respect to total concentration, freezing point and osmotic pressure. With respect to the relative concentrations of the inorganic cations, there are often striking similarities. The composition depends on the habitat of any particular species. In marine invertebrates such as the lobster, the crab, the scallop and the squid, the composition of the circulating fluid resembles that of the environmental sea water, which is the *milieu intérieur.* These species have not acquired the independence of the internal fluids from the external environment, which BERNARD considered to be the criterion for a free and independent life. The frog, which is a fresh water amphibian, has a circulating fluid which bears little resemblance to that of the external environment, with regard to electrolyte composition. Like many other amphibians and fresh water vertebrates, evolution has endowed the frog with a certain freedom from the external environment. The freezing point and osmotic pressure of frog blood plasma are much different from the values for the fresh water habitat. The values correspond to those for an approximately 0.7-percent solution of sodium chloride. In comparison, mammalian blood plasma is isotonic with a 0.9-percent solution of the same salt. Frezying point depression and osmotic pressure are approximately proportional to the salt concentration.

Sea Water

The total electrolyte concentration of sea water in the present geological age is of the order of three times that of mammalian blood plasma or of Ringer's solution. The freezing point depression is also about three times as great. Marine invertebrates are adapted to a cosmic environment that is greatly different from that of fresh water species. In many cases, the composition of the blood plasma of marine invertebrates resembles that of sea water more closely than it resembles that of marine vertebrates such as fish. MACALLUM [1910] investigated the subject, and concluded that the electrolyte composition of the blood plasma of all vertebrates is derived from the sea water in which the ancestral species flourished and developed. In his words: 'Enough, however, has been advanced here to make it extremely probable that the inorganic composition of the blood plasma of vertebrates is an heirloom of life in the primeval ocean.' MACALLUM based this conclusion on analytical data for the electrolyte composition of vertebrate blood plasma as compared with the composition of sea water. 'So, from the composition of the blood plasma in vertebrates, we may infer the relative composition of the ocean in the remote past, when the vertebrate kidney acquired the function of controlling the salts and their concentration in the blood. This would give for the ocean of that age the ratios in round values as follows:

Na, g	K, g	Ca, g	Mg, g
100	6.00	3.00	1.50–2.00

Compare this with ratios of the ocean of today, which are:

Na, g	K, g	Ca, g	Mg, g
100	3.613	1.912	12.106

It is instructive to express these figures on the basis of moles of each cation per 100 mol of sodium. This yields:

	Na	K	Ca	Mg
Blood plasma	100	3.41	1.66	1.50–2.00
Sea water	100	2.31	2.25	11.64

'This would place the origin of the distinctively vertebrate type, that is, of a form possessing a kidney with a function like that of the vertebrate excretory organ at a time when the ocean had one eighth of the present concentration of magnesium, and less than one third of its present concentration of sodium chloride.'

Table I. Composition of the serum of various vertebrates[1]

Species	Na	K	Cl	Ca	Mg
Man	135–145	4.0–5.5	96–107	4.5–5.5	1.1–2.1
Pig	144	6.0	106	5.6	2.2
Rabbit	140	5.1	102	5.6	2.2
Rat	142	5.5	110	–	3.0
Ave., mEq	141.5	5.3	105	5.4	2.25
Ref. to Na	100.0	3.74	74	3.82	1.58

1 WIDDOWSON and DICKERSON [1964].

Based on more recent analytical data, calculations of the relative ion concentrations in blood plasma of mammals and marine vertebrates and of sea water may be compared with MACALLUM's earlier figures. Table I shows results for four species of mammals (man, pig, rabbit and rat) on the basis of 100 mM sodium.

	Na	K	Ca	Mg	Cl
Mammalian plasma	100	3.74	1.91	0.79	74

With the exception of Mg, the figures for the plasma cations are in good agreement with MACALLUM's results. Data for the sea water habitats of four species are given in table II. They yield the following results on a molal basis, as referred to sodium:

	Na	Ka	Ca	Mg	Cl
Sea water	100	2.11	2.23	11.4	117

Again, the results agree well with those of MACALLUM. The results for potassium, calcium and magnesium agree within 1 or 2% with his figures.

Recent data for the plasma of marine invertebrates are compared in table III. Calculations for the molal concentrations per 100 mol sodium indicate:

	Na	K	Ca	Mg	Cl
Vertebrates	100	1.45	1.55	1.50	83
Invertebrates	100	2.33	2.53	2.23	108

Table II. Electrolyte composition of sea water in various habitats (mM/l)

Species	Na	K	Ca	Mg	Cl
Hagfish[5]	506	10.8	11.1	57.8	592
Roman eel[1]	564	12.0	12.4	64.3	659
Lobster[2]	457	9.8	10.1	52.2	535
Sea urchin[3]	485	10.0	11.4	55.0	566
Average	503	10.6	11.2	57.3	588
Ref. to Na	100	2.11	2.23	11.4	117

1 ROBERTSON [1960].
2 ROBERTSON [1961].
3 ROTHSCHILD and BARNES [1953].

Table III. Electrolyte composition of blood serum

Species	Na	K	Ca	Mg	Cl
Cold-blooded vertebrates					
Eel, *Anguilla anguilla*[1]	183	3.2	2.4	3.8	155
Roman eel[2]	212	1.95	3.9	2.2	188
Salmon[3]	212	3.6	–	–	157
Average	202	2.9	3.1	3.0	167
Ref. to Na	100	1.45	1.55	1.50	83
Invertebrates					
Lobster[4]	517	8.6	16.2	10.4	527
Crab *Carcinus maenas*[5]	468	12.1	8.7	11.7	524
Mytilus edulis[6]	502	12.7	–	–	525
Pecten[7]	490	12.5	–	–	573
Average	494	11.5	12.5	11.0	542
Ref. to Na	100	2.33	2.53	2.23	108

1 Chan *et al.* [1967]
2 Robertson [1960]
3 Parry [1961]
4 Robertson [1961]
5 Shaw [1955, 1958a, b]
6 Potts [1955]
7 Potts [1958]

When compared with sea water, it is evident that the greatest difference occurs with magnesium, when plasma is compared with the *milieu intérieur*. The external ratio of sodium to magnesium is 100/11.4, or about 9 to 1. The internal ratios for vertebrates and invertebrates are, respectively, 100/1.5 and 100/2.23. For mammals, the ratio is 100/0.79, or about 125. This is about double the ratio for marine vertebrates. In all cases, the plasma concentration of magnesium is much less than the external concentration in sea water, when the values are expressed in relation to sodium.

The total ion concentration in the plasma of marine vertebrates usually corresponds to a freezing point of about −0.6 to −0.8 °C [Love, 1970]. The freezing point of sea water is usually of the order of −1.6 to −1.8 °C. Thus the external salt concentration may be two or three times as great as the internal concentration, as the freezing points differ by 1.0 °C or more. Marine invertebrates differ from the vertebrates in this respect. The ion concentrations of plasma are much higher, and the freezing points of invertebrate plasma approach those of the external environment.

The adaptations of fish and other aquatic species to their cosmic environment have been achieved through evolution of the kidney. The conservation of the *milieu intérieur* of the higher vertebrates, including mammals, required the development of a special type of kidney. All the gross anatomical and physiological modifications through which the higher vertebrates have evolved, have been acquired under the condition of a circulating fluid of almost invariant electrolyte content (table I).

The Kidney[1]

The kidney is a glandular organ of widely varied structure in different species. Two functions may be distinguished. One is osmoregulation, or control of the electrolyte concentration of the blood. The other is secretory, serving to dispose of the waste products such as urea and other nitrogenous substances. In marine invertebrates, the electrolyte concentration of the circulating fluid is very similar to that of the environmental sea water. The problem of osmoregulation is relatively simple, and the kidney function is mainly secretory. In the evolution of the vertebrate kidney from the primitive type, a new function was assumed. This was to excrete water in large quantities in the urine, retaining the blood electrolytes. This function was developed in fresh water vertebrates, and was retained in the evolutionary transition to terrestrial forms. The vertebrate kidney regulates the water and salt content of the blood. In addition, it excretes the waste products.

From the structure of the kidney of the primitive fishes, it is inferred that the kidney of the early protovertebrates was segmented. Each segment of the body is thought to have carried a bilateral pair of nephric tubules, each of which communicated with the body cavity (coelom) through an open mouthed coelostome. The function of the renal tubule was to carry off the fluid from the coelom. The fluid was excreted through the body wall by drainage through the tubules. In the evolution of higher vertebrates, a glomerulus was developed. In the earlier stages, the glomerulus fitted loosely into the coelostome, but in later stages of evolution, it was fused within the end of the tubule. In the typical nephron of higher vertebrates, the coelostome no longer exists. In the fully evolved kidney of mammals, the tubules are supplied by the post-glomerular blood of the renal circu-

1 MARSHALL [1934], RICHARDS [1938].

lation. The kidneys of fishes, amphibia, reptiles and birds retain a more primitive type of circulation in which the tubules receive blood through the 'renal-portal' circulation.

The glomeruli of the kidney were discovered by MALPIGHI in 1666. They were described as tufts of capillaries. MALPIGHI also described the capillary circulation of the lungs of frogs. These discoveries occurred a few decades after HARVEY's demonstration of the circulation of the blood (1628). By explaining the movement of blood from the arterioles to the venules, MALPIGHI's observations fully supplemented HARVEY's demonstration. Modern ideas as to the function of the glomeruli developed historically through the following steps.

1. The description of BOWMAN's capsule (1842), a space which connects the glomerulus, or 'malpighian corpuscle' with the lumen of the tubule.
2. LUDWIG's evidence of passive filtration of water, electrolytes and waste products through the glomerulus (1844).
3. Support and confirmation of LUDWIG's ideas by STARLING and by CUSHNY (from about 1900 to 1917, the date of publication of an important book by CUSHNY).
4. Direct proof of the Ludwig-Starling-Cushny theory by RICHARDS and his collaborators [1938]. Direct analysis of the glomerular filtrate.
5. The demonstration of tubular excretion by MARSHALL [1934]. Studies on the aglomerular marine fishes, in which urine formation occurs through tubular excretion.

Present day studies on renal physiology are concerned with the rates and dynamics of the glomerular filtration and with the processes of tubular excretion and reabsorption. In the tubules, water and the blood electrolytes are reabsorbed to the extent required to maintain a constant blood volume and constant electrolyte concentrations. The maintenance of the proper acid-base balance of the blood is also a function of the tubules. This is managed by regulation of the excretion of chloride, ammonium ions, non-protein nitrogen and hydrogen ions. The lungs cooperate in the regulation of blood pH by the control of carbon dioxide tension in the alveoli and in the blood.

The invariance of the *milieu intérieur* over the entire course of vertebrate evolution depends on the unique development of the kidney, occurring together with a parallel evolution of the respiratory and circulatory systems. This was a necessary condition for the development of the musculature, the nervous system and the skeleton. The evolution of the species demands the development of specialized organs such as the kidney and the lungs.

It also requires concurrent processes of biochemical evolution. The respiratory functions of the blood depend on the special properties of unique substances such as the hemoglobins of mammals or the hemocyanins of marine invertebrates. Unique physiological and biochemical adaptations are thus superimposed on the most primitive properties of the *milieu intérieur,* common to the body fluids of vertebrates and invertebrates in all the stages of their evolution. The primitive properties are those that are common to Ringer's solution and the ancestral sea water of remote paleozoic ages. These fluids are described in terms of the physicochemical nature of simple inorganic cations and anions, and of water itself. Wherever they occur, whether in the *milieu extérieur,* the *milieu intérieur,* or in the cells and tissues of the body, sodium, potassium and chloride ions retain certain elementary properties which depend on their positions in the periodic table of the elements. These are measurable quantities which determine their physicochemical behavior, not only in water, but also, we must assume, in any material physicochemical system, whether or not it is living.

The Electrolytes

The *milieu intérieur* of mammals contains in solution the following inorganic ions: sodium, potassium, calcium, magnesium, chloride, sulfate, bicarbonate and phosphate. It is convenient to count only the first five as the normal inorganic physiological ions. Phosphate ions enter into many metabolic reactions that involve the formation of their esters. They also react with calcium ions to form insoluble bone salts, with which they are in equilibrium. Unlike the four physiological cations, bicarbonate ions are produced in cellular respiration, and are transported irreversibly to the external air in the form of carbon dioxide. Sulfate ions are present in blood plasma only at very low concentrations; they react with extracellular substances to form esters such as chondroitin sulfate. In contrast with sulfate, phosphate and bicarbonate ions, sodium, potassium, calcium, magnesium and chloride ions do not tend to esterify or to enter into irreversible metabolic reactions. Calcium and phosphate react to form insoluble salts in bone and teeth, but the salts are in reversible equilibrium with the ions of the *milieu intérieur*.

With respect to their positions in the periodic table of elements, the five physiological ions are closely related. As a series in the order of increasing atomic weight and atomic number, they may be arranged as follows:

	Element				
	Na	Mg	Cl	K	Ca
Atomic Number	11	12	17	19	20

The ionic forms of sodium and magnesium thus have the electronic configuration of the inert gas, neon (atomic number 10). Chloride, potassium and calcium ions are characterized by eight additional electrons. They are larger than the other two ions, and conform to the argon type of structure (atomic number 18). In the periodic table, the most stable atomic structures up to atomic number 18 are those of helium (2), neon (10), and argon (18). Lithium ionizes to form the helium configuration, whereas sodium and potassium ions resemble the neon and argon cores. Thus the four inorganic cations are incapable within the animal organism of forming stable covalent or non-ionic bonds. Chloride ions, on the other hand, can form covalent bonds in carbon compounds, but for the most part, it may be assumed that chloride reacts reversibly within the animal organism to form only weak ionic bonds or electrostatically stabilized complex aggregates.

Since about 1900, most physical chemists have regarded salts such as NaCl, KCl, $CaCl_2$ and $MgCl_2$ to be completely dissociated in dilute aqueous solution. In blood plasma, about half the calcium and half the magnesium have been considered to be non-dialyzable; these fractions are bound to the plasma proteins in non-ionized forms. The equilibrium between ionized and bound forms is reversible. Only the ionized form is physiologically active; it has been studied through its effects on the irritability of heart muscle, for example. Since the inorganic anions and cations form no undissociated complexes with each other, their physiological properties and effects of any ion may be regarded as independent of the other ions [Hofmeister, 1888; Loeb, 1906]. Each of the ions may be regarded as independent of the other ions [Hofmeister, 1888; Loeb, 1906]. Each of the ions may be regarded as manifesting permanently invariant properties that depend only on its inherent character, as determined by its atomic number, or position in the periodic table. Among other properties, this determines its stable radius, which is a periodic characteristic. The radius of each kind of ion may be determined from studies of crystal structure. In any kind of crystal (NaCl, KCl, $CaCl_2$...), the radius of either ion is independent of the nature of the other ion [Goldschmidt, 1926; Pauling, 1944].

The measured radius of the Cl ion, for example, is independent of the nature of the crystal. This property resembles many other properties of ions. Each kind of ion in solution is characterized by its own specific conductance, its own ionic mobility, polarizability, refractive index, and so on. What is true of electrolyte solutions must also be true of biological systems. According to the theory of electrolytic dissociation, each ion has a set of properties that are independent of the nature of the other ions that are present. This leads to an independence of their biological properties. This was recognized in the work of early investigators such as RINGER [1880–1882], HOFMEISTER [1888], LOEB [1906], HARDY [1905], MELLANBY [1905], and MACALLUM [1910].

The crystal radii of the inorganic ions thus represent only one of a set of important properties that depend on the atomic structure and on the atomic number of the element. GOLDSCHMIDT's values for crystal radii, as summarized by PAULING, are given in table IV. Important thermodynamic properties of the ions, both *in vivo* and *in vitro,* are related to the crystal radii [JOSEPH, 1971a]. Since many other non-thermodynamic properties (mobilities, polarizabilities, and so forth) are also based on atomic structure and atomic number, they should also be considered to have a bearing on the fundamental biological properties of the *milieu intérieur*.

All such properties should be considered to be invariant on the geological time scale, and from the point of view of phylogeny. They are the conservative or non-evolving properties, tending to be independent of genetic mutations. These are the properties which, acting together with the stability of the genes and chromosomes, make possible the enduring patterns of life within a given genus or species. 'Conservation of reversibility' depends essentially on the independence of geological time, that is shown by any species with a very slow rate of evolution.

Dielectric Properties of the Internal Environment

Pure liquid water is a solvent with a temperature dependent dielectric constant of about 78–81 [DEBYE, 1929; COHN and EDSALL, 1944]. For the purposes of subsequent calculations, a value of 80 at temperatures from 18–37 °C may be assumed without serious error. The calculated value for mammalian blood plasma turns out to be of the order of 110. The plasma globulins have 'dielectric increments' of about 1.1 units/g protein/l. The globulin content of human plasma is about 29 g/ l. Then the calculated dielectric constant is:

$$D_{calc.} = D^\circ + 1.1 \times 29 \tag{2.1}$$

where D° is the dielectric constant of water. If this is taken as 80, the calculated value of the dielectric constant of blood plasma is approximately 112.

In the theory of electrolyte solutions, certain important properties of the ions depend on ionic charge, z_i, ionic radius, b_i, and the dielectric constant D. The molal electrostatic work, w_i of charging a system of N_i ions in a medium of dielectric D, is calculated as:

$$W_i = \frac{N_i z_i^2 e^2}{2\,Dr_i} \tag{2.2}$$

where N is Avogadro's number (6.02×10^{23}), and where e is the unit of electrostatic charge (4.8×10^{-10} ESU). If the dielectric constant is assumed to be 80, as for pure water:

$$W_i = \frac{164\,z_i^2}{D\,r_i}\ (\text{kcal}) \tag{2.3}$$

where r_i is the crystal radius as given in table IV. According to LAIDLER and PEGIS [1957], it is necessary to correct this value for ions in solution. The uncorrected value of r_i, as used by BORN [1920], had yielded values of W_i that were considerably higher than observed values. In an aqueous dielectric:

$$W_i = \frac{164\,z_i^2}{D\,b_i} \tag{2.4}$$

Table IV. Crystal radii of ions[1]

O^{--}	F^-	Na^+	Mg^{++}	Al^{3+}	Si^{4+}
1.32	1.33	0.98	0.78	0.57	0.39
S^{--}	Cl^-	K^+	Ca^{++}	Sc^{3+}	Tl^{4+}
1.74	1.81	1.33	1.06	0.83	0.64
Se^{--}	Br^-	Rb^+	Sr^{++}	Y^{3+}	Zr^{4+}
1.91	1.96	1.49	1.27	1.06	0.87
Te^{--}	I^-	Cs^+	Ba^{++}	La^{3+}	Ce^{4+}
2.11	2.20	1.65	1.43	1.22	1.02

1 Values in Å [GOLDSCHMIDT, 1926; PAULING, 1944].

where b_i is the corrected value of r, obtained by multiplying by the factor 1.25. Then for sodium ion, b_i is approximately 1.25 Ångstrom units. When D is 80,

$$W_i = 1.64 \text{ kcal} \tag{2.5}$$

This represents the value of $S\Delta\mu_{Na}{}^{\circ}$, the *standard chemical potential* of sodium ions in water or blood plasma. The values of the standard chemical potentials of the other ions are given by related formulae:

	Ion Na	K	Mg	Ca	Cl
Standard chemical potentials in kcal	1.64	1.23	8.2	6.15	0.90

Each of these values is proportional to z_i^2/b_i. According to table IV, the values of b_i increase in the series: Mg (0.78); Na (0.98); Ca (1.06); K (1.33); Cl (1.81).

The radius of magnesium ion is about eight tenths that of sodium, and the ratio of calcium radius to that of potassium is likewise eight tenths. The radius of chloride is about 1.8 times that of sodium. Therefore, referred to sodium, the values of $\mu_i{}^{\circ}$ are approximately as follows:

Cl	K	Na	Ca	Mg
0.55	0.75	1.0	3.75	5.0

For any ion these values are proportional to z_i^2/b_i. The *change* of standard chemical potential of any ion between any two phases is important in determining ion distributions in a heterogeneous system. For any ion:

$$\mu_i = \mu_i{}^{\circ} + RT \ln c_i f_i \tag{2.6}$$

where μ_i is the chemical potential, $\mu_i{}^{\circ}$ is the standard chemical potential, c_i is the concentration in a given phase, and f_i is the activity coefficient [Lewis and Randall, 1961]. In ordinary salt solutions as studied *in vitro,* f_i is taken as 1.0 in the reference state at infinite dilution. This procedure is applicable only to true salt solutions in which the ion concentrations can be varied continuously from infinite dilution to any desired concentration, c_i.

Biological systems cannot be treated in the same way as the unrestricted systems of *in vitro* solutions. Living cells and tissues are always constrained with respect to the variability of the *milieu intérieur* and of all other kinds of phases. The standard states of maximal accessibility may be treated as invariant; stimulated or normally accessible states represent an increase of degrees of freedom, in which variations of the concentrations are small and constrained. Cells and tissues never approach states of infinite dilution. Therefore, the most logical procedure for this type of invariant system is to define the standard physiological state as the reference state, in which all activity coefficients are assigned the value of 1.0 by definition [JOSEPH *et al.*, 1961, 1964, 1965; JOSEPH, 1971a, b]. Accordingly, the configurational term in the chemical potential is given by $\Delta\mu_i^\circ$, as in the following equation:

$$\Delta\mu_i = \Delta\mu_i^\circ + RT \ln \frac{c_i''}{c_i'} \tag{2.7}$$

Here $\Delta\mu_i$ denotes the change of chemical potential between two phases, and $\Delta\mu_i^\circ$ is the change of standard chemical potential. R is the gas constant (1.987 cal/mol/deg.), and T is the absolute temperature. The concentrations of free ions in the two phases are represented by c_i'' and c_i'. In subsequent chapters, it will be shown that the electrical potentials and distributions of the five physiological ions in biological systems depend on the values of $\Delta\mu_i^\circ$. Therefore, they depend on the dielectric constant of each kind of phase and on the charge and radius of each kind of ion.

In applying these formulae, the *milieu intérieur* is treated as a homogeneous fluid, the dielectric constant of which is about 80. It has been found that blood plasma and the aqueous phases of the *milieu intérieur* have very similar, approximately constant, electrolyte compositions. Fluids such as aqueous humor, synovial fluid and edema fluid have approximately the same electrolyte composition as aqueous dialysates of blood plasma [ROPES and BAUER, 1953; JOSEPH, 1971b]. Therefore, blood plasma is equivalent thermodynamically to a protein-free dialysate or to an isotonic *in vitro* protein-free electrolyte solution. This is theoretical justification for treating plasma and the other liquid phases as fluids with dielectric constants of about 80. It is then unnecessary to consider the effect of the plasma proteins, which elevate D to a level of about 110. The value of μ_i° in any intracellular phase appears to be determined by the state of aggregation of the n macromolecular components, which in turn determines the state of aggregation and dielectric properties of the intracellular water.

'A solution is defined as a homogeneous mixture, which can pass through variations of composition within the limits that define its existence' [ADAMS, 1919]. According to this definition, only blood plasma and other homogeneous liquid phases of the body can be described as solutions. In homeostasis, however, these fluids are very nearly invariant; they must be distinguished from true *in vitro* solutions of the same components, which would have many degrees of freedom. In the living organism, the composition of blood plasma is restricted by the same thermodynamic conditions of constraint that apply to the organism as a whole. The composition nearly always approximates that of the standard state of maximal stability.

Cells and connective tissues can never be considered to be homogeneous solutions. They are heterogeneous structures in which water must be considered to exist in two or more states. A uni-state theory of intracellular water necessarily implies a physicochemical state of homogeneity. According to ADAMS' definition of a solution, intracellular phases would be homogeneous and in the same state of aggregation as all other liquid phases. This is inconsonant with GIBBS' definition of 'phase' [GIBBS, 1928; JOSEPH, 1971a]. Such homogeneous systems would have no negative configurational entropy and no positive configurational free energy. A uni-state theory of water structure eliminates any possibility of intracellular changes of state. The values of intracellular and extracellular dielectric constant would be the same, and the values of $\Delta\mu_i^\circ$ of all the ions would approach zero.

The classical 'Donnan equilibrium' or 'membrane equilibrium' implies such a uni-state theory of the aqueous dielectric. In its present state, the theory of active transport requires mechanisms of ion transport that depend on metabolic processes which generate energy in the membrane. Then, according to chapter I, it is implied that M, Permeability $\supset$ T, A, B, C...

The uni-state theory therefore requires the primary properties which describe intracellular morphology, composition and state to be determined by secondary properties of the cell, including metabolism and membrane permeability. All the other secondary properties such as resting and action potentials, muscular work and tension, would then logically depend on secondary properties. A uni-state theory of intracellular morphology therefore eliminates all dielectric properties of the cell (since D is constant). It would also eliminate changes of the chemical and electrical potentials of ions and of configurational entropy and free energy. According to SCHRÖDINGER [1944], these properties depend on the organization of intracellular macromolecules as well-ordered aggregates maintained by nutritional sources of negative entropy.

Such systems are necessarily heterogeneous aggregates that consist of two or more well-ordered states. Changes of state imply changes of free energy, entropy, chemical and electrical potentials. This contrasts with the results of the existing uni-state theory, which by its nature explains the energetics of physiological responses as changes of chemical bond energy or enthalpy rather than as univariant changes of physicochemical state. A uni-state theory tends to be mechanistic rather than thermodynamic. It therefore tends to lead to the 'fallacy of misplaced concreteness' [WHITEHEAD, 1925].

Chapter III: Protoplasm

'It may seem a small thing to admit that the dull vital actions of a fungus or a foraminifer are the properties of their protoplasm, and are the direct results of the matter of which they are composed. But if, as I have endeavoured to prove to you, their protoplasm is essentially identical with, and most readily converted into, that of any animal, I can discover no logical halting place between the admission that such is the case and the further concession that all vital action may, with equal propriety, be said to be the result of the molecular forces of the protoplasm which displays it. And if so, it must be true, in the same sense and to the same extent, that the thoughts to which I am now giving utterance, and your thoughts regarding them, are the expression of molecular changes in that matter of life which is the source of our other vital phenomena' [HUXLEY, 1868].

The term 'protoplasm' seems to have first been used by PURKINJE [1839] to denote what is first formed in the plant and animal cell. A few years later, v. MOHL [1846] working on plant cells, distinguished not only the cell wall and the nucleus from the cell contents, but also the tough, slimy, granular semi-fluid constituent from the watery cell sap generally confused with it under the common name. He proposed 'protoplasma' from *προτοσ* ('first') and *πλασμα* ('formed substance'). This definition was apparently independent of PURKINJE's term, and it had a chemically oriented meaning for a nitrogenous material which was affected by alcohol and acids, and stained by iodine.

Microscopic observations of the form and movement of living protoplasm may be dated from the middle of the 18th century. The amoeba was studied by v. ROSENHOF [1755]. Rotation and other movements were seen in plant cells by CORTI [1774], MAYEN [1827] and BROWN [1821]. Before this period, plant and animal 'cells' were held to be different. During the 19th century, random and oscillatory and rotary movements, observed by BROWN (Brownian movement) were subsequently explained on many different grounds: consequently there was no explanation at all. The true explanation was finally given in a series of papers by EINSTEIN [1926]: it is based on the kinetic theory of matter applied to colloidal particles or molecular kinetic units of high molecular weight. The kinetic energy of the particle is $\frac{1}{2}mv^2$, and the random, diffuse motion observable under the microscope may be so explained. Subsequent application of the theory of Brownian movement to the results of PERRIN and his students led to estimates of Avogadro's number. The studies of SVEDBERG [1928] on sedimentation velocity and equilibrium are based on the same principles.

It may be inferred from the observations of BROWN and the studies of EINSTEIN, PERRIN and SVEDBERG that the protoplasmic substances or life slimes are in no way exempt from the general laws of thermodynamics or kinetic theory, and that under the microscope they behave like all other colloidal substances. It may also be inferred that the molecular kinetic unit is very large, and that the molecular weight is much greater than that of ordinary albumins and globulins. In the series of papers by EINSTEIN, the theory of the Brownian movement, of the osmotic pressure and of the diffusion of macromolecular substances was based on the classical theories of heat and thermodynamics. Any physicochemical theory of the life slimes must take into consideration these general principles of physical chemistry.

The early observations of v. ROSENHOF, CORTI, MAYEN and BROWN culminated in the cell theory of SCHLEIDEN and SCHWANN [1838–1839]. This theory, however, contained views on the structure, origins and functions of the cell substances which proved to be erroneous and misleading. The work of DUJARDIN [1835] was essentially a reversion to the point of view of v. ROSENHOF. Studying foraminifera, he wrote: 'Je propose de nommer ainsi (sarcode) ce que d'autres observateurs ont appelé une gelée vivant, cette substance glutineuse, diaphane, insoluble dans l'eau, se contractante en masses globuleuses, s'attachant aux aiguilles de dissection, et se laissant étirer comme ce mucus, enfin se trouvant dans tous les animaux inférieurs interposée aux autres éléments de structure.' He recognized 'sarcode' to be present in polyps, ova of the slug, and in other cellular material.

ECKER [1849] compared the formed contractile substance of muscle with the unformed contractile substance of lower organisms. Shortly after, DONDERS referred contractility of the cell membrane to the actual cell substance. COHN, about this time proposed that vegetable protoplasm and animal sarcode, if not identical, must at any rate be highly analogous substances. SCHULTZE [1854] seems to have finally identified vegetable 'protoplasm' with animal 'sarcode'. He replaced the early concept of the cell based on an all important membrane enclosing a nucleus surrounded with fluid by the concept of a unit mass of living matter or protoplasm.

Our present usage of 'protoplasm' for the living substance of the animal as well as the plant may be considered to date from SCHULZE [1861]. Notwithstanding DUJARDIN's priority, the term 'sarcode' had lapsed by 1885, at the latest. The state of opinion at the beginning of the 20th century may be indicated by a quotation from MINCHIN's article in Encyclopaedia Britannica (1902). 'Competent microscopists assert that granules of protoplasm do not lie in a structureless matrix but are embedded in the substance of a minute and delicate framework or morphoplasm which in its turn is bathed by a watery fluid or enchylema permeating the whole substance.' The 'minute and delicate framework or morphoplasm' was considered possibly to be either a tangle of separate fibrillae or a reticulum or meshwork with fine meshes ($\frac{1}{2}$ to 1 μm). Any theory would be required to account for solid states of aggregation. At about this time BÜTSCHLI [1892] proposed a honeycomb or alveolar structure similar to a foam in which the interstices are filled with another liquid, i.e. a fine emulsion of two immiscible liquids. BÜTSCHLI's theory is monomorphic. FISCHER's [1894–1895] concept is that of a polymorphic substance, which may be granular, reticular, alveolar or homogeneous at different times. The fundamental units are proteins. Precipitated in different ways, they assume different histological and staining forms.

More than half a century's development has led to a far deeper understanding of the submicroscopic morphology of the intracellular cytoplasm and the extracellular phases of connective tissue. This is based on both physical and biochemical results. The structure and organization of biological macromolecules and of numerous kinds of complex aggregates has been studied by optical methods (birefringence, absorption spectra, refractometry), X-ray diffraction studies of protein or polysaccharide fibers, and by the application of techniques of electron microscopy. A recent authoritative view is that of FREY-WYSSLING [1953]: 'The substratum in which life is inherent is not a disperse phase with individual particles or ultramicrons; it possesses a *structure*. Its active centers, which control development, are arranged in a given order. They are not intermingled by mere laws of chance and brownian molecular movement; the fact is rather that they arrange themselves into a delicate, very plastic and flexible pattern, actuated, as it were, by a purposeful, coordinated impulse . . . For, orderly biological processes are unthinkable without presupposing structure, and it is therefore out of the question that any living constituent of protoplasm could consist of structureless, fluid, independently displaceable particles.'

The briefest survey of studies on protoplasm and the life slimes is sufficient to show the nature of the problems that are involved. The organization of the cell depends on the properties of the macromolecular substances, the intermolecular forces which result in definite states of aggregation, and the special characters of the liquid and solid phases leading to observable physicochemical and biological properties. These are the properties of proteins, conjugated proteins and colloidal coacervates; ultimately they are related to the chemical structure of the macromolecular and polyelectrolyte components. The group of life slimes with common properties, by a slight extension of definitions, may be extended to include the extracellular substances of the surrounding colloidal matrices. This classification would then include the mucoproteins, mucopolysaccharides, glycoproteins and fibrous proteins such as collagen, reticulin and elastin. This group includes polyelectrolytes of high molecular weight, which may be classified as soluble or insoluble, isotropic or anisotropic. These substances all tend to form complex aggregates and to exist in various possible states of aggregation. The limits of these states with respect to composition define the range of possible physicochemical states.

All theories seem to indicate a heterogeneous (liquid-solid) intracellular structure. No theory seems to imply that either state is well-defined or homogeneous. The liquid state is apparently not one of high fluidity, disorder or entropy. Nor is the solid state one of high rigidity, order and low entropy, like that found in symmetrical crystals. The behavior of the heterogeneous system seems to be that of one near a critical state or a transition point, where the properties of the phases tend to converge. This condition might imply that the attractive forces which stabilize the aggregates include thermolabile von der Waals forces of cohesion: dipole-dipole, dipole-induced dipole, and the London dispersion forces of attraction between oscillating systems of charges. In addition, hydrogen bonds between electronegative oxygen and nitrogen atoms should be included. Attractions of this kind are thermolabile and of low energy; many kinds of rearrangement would be possible. The order in the system would be high, the entropy would be low, and responses to stimuli would be sensitive. The recognized physiological properties of protoplasm include those of growth, reproduction, and what is called 'irritability'. These properties must ultimately be referred to the nature of the components of the colloidal aggregates and to the internal stabilizing attractions involving water and other substances of low molecular weight, including the inorganic ions.

Nucleoproteins

The nucleic acids (RNA and DNA) are intracellular components. In varying quantities they are found in all cells of plants and animals, occurring in the cytoplasm as well as in the nucleus. The nucleic acid of the cytoplasm of both plant and animal cells is ribonucleic acid (RNA). That of the cellular nuclei is deoxyribonucleic acid (DNA). Abundant quantities of nucleic acid occur in spermatazoa, oocytes, and in the thymus gland. Somatic cells usually contain only small quantities. From most cells nucleic acids are isolated only with great difficulty. In general, they are bound to protamines, histones and other intracellular proteins. The process of isolating the components of these aggregates may result in extensive degradation of all the products. Combinations of nucleic acid with proteins have been classed as nucleoproteins, but these products are complex colloidal aggregates, the properties of which depend as much on the method of preparation as on the nature of the native material.

The modern phase of research on the nucleic acids, protamines and histones dates from work of MIESCHER [1874] on the biology of the Rhine salmon *(Salmo salmar)*. MIESCHER, working at Basel, was able to study the salmon during their journey from the North Sea to their spawning beds in the upper Rhine. In this phase, the fish consume no food. There is a considerable loss of body weight. An extensive loss of muscle tissue is accompanied by a very great increase in the mass of the reproductive organs. It was concluded that the production of eggs and spermatazoa occurs through the conversion of the muscle proteins to nuclear material. Not all the protein is so converted; a large quantity is required to support the energy metabolism of the salmon during this period, when production far exceeds consumption.

During the spawning season, the spermatazoa are available in very large quantities. They are composed of heads, tails and middles, but the heads contain by far the greater proportion of the total mass. By treatment with acetic acid, the tails are dissolved, and the heads remain as granular colloids. The head of the spermatazoon is regarded as a metamorphized nucleus. MIESCHER found the heads to be free of the 'higher' proteins, and to consist almost exclusively of an organic nitrogenous base and an organic acid which contains phosphorus. He clearly recognized that his protein-free organic acid was not a cleavage product of protein, and that his nitrogenous base was not an ordinary protein. The acid is now classified as a nucleic acid, and the base as salmine, a protamine. The discovery has been of great importance in the following respects:

1. It inaugurated the chemistry of nucleic acids and its relation to the chemistry of purines, pyrimidines, carbohydrates and esters of phosphoric acid.

2. It established the biological importance of the protamines and histones, which are basic proteins that are rich in one or more of the basic amino acids: arginine, lysine and histidine, and with low contents of the other amino acids.

3. It clearly showed that nucleic acids, protamines and histones are synthesized from protein.

4. By localizing the nucleic acid in the heads of spermatazoa, it pointed to the importance of these substances in the transmission of hereditary characteristics.

Nucleic Acids (RNA and DNA)

Ribonucleic acid was studied chemically by Kossel [1892]. In that year Kossel and Neumann studied the products obtained by the hydrolysis of deoxyribonucleic acid (DNA). During the following years, Kossel and his associates carried out extensive studies of the nucleic acids, identifying the hydrolytic products.

Each nucleic acid contains four nitrogenous bases (purines and pyrimidines), a pentose, and phosphoric acid. These substances are the ultimate products of hydrolytic cleavage of the nucleic acid. The intermediate products are classified as polynucleotides, mononucleotides and nucleosides. The mononucleotides and polynucleotides contain esterified phosphoric acid groups. The nucleosides contain a purine or pyrimidine base conjugated with a pentose. The molecular weights of the mononucleotides are approximately 360–385, as is shown by the composition: H_3PO_4, 99 g; d-ribose, 150 g; pyrimidines, 111–126 g; purines, 135 g.

Since there are both primary and secondary phosphate groups, the equivalent weights of mononucleotides are low (about 180–190 g/Eq). Polymerization occurs through the phosphate groups, so that many of the secondary groups in the nucleic acids are not titrable. The equivalent weights are therefore of the order of 400 g/Eq. The values are inversely proportional to the number of titrable phosphate groups per gram of nucleic acid or polynucleotide.

Extracellular Substances

It is not always possible to make a precise distinction between intracellular and extracellular macromolecular components. If all chemical synthesis is intracellular, then the extracellular substances originate from metabolic processes within cells. Nevertheless, there are definite distinctions between the two types of component. For example, the soluble globular type of protein, including the serum albumins and globulins, are mainly extracellular. The nucleoproteins and nucleohistones are intracellular. Connective tissues are rich in glycoproteins, mucoproteins and mucopolysaccharides. These contain carbohydrates, and are important components of the amorphous ground substance and of the basement membrane. Hyaluronic acid and chondroitin sulfuric acid are highly polymerized products which are built up of hexosamine and glucoronic acid residues. They occur as components of complex aggregates which contain glycoproteins, mucoproteins and collagen. The extracellular components may contain considerable quantities of hexosamine and glucuronic acid residues. Pentoses, on the other hand occur intracellularly as constituents of RNA and DNA.

Extracellular and intracellular polymers have low equivalent weights. For hyaluronic acid, 397 g would be required to neutralize 1 Eq of NaOH. The equivalent weight of the acid may then be taken as 397 g. This is estimated on the basis of one glucuronic acid residue alternating with one acetyl glucosamine residue. Chondroitin sulfuric acid has a lower equivalent weight because, in addition to the glucuronic acid residue, there is an esterified sulfate group. The equivalent weight is 239. The chondroitin sulfates are polymers of acetyl galactosamine and glucuronic acid. Hexosamines are of widespread occurrence in the animal kingdom. The hard integument of lobsters, crabs and insects is composed of chitin, a polymerized form of acetyl glucosamine. X-ray diffraction studies show a structure similar to that of cellulose. Polymerized forms of glucuronic acid occur widely in plant tissues. In animal tissues, the polymers contain, in addition, some form of hexosamine. Cellulose, the important structural fiber of plants, is of very rare occurrence in the animal kingdom. It is found, for example in the mantle of tunicates. In earlier times, starches were considered to occur only in plants. This belief was dispelled by BERNARD's discovery of glycogen synthesis in the liver. In the tissues, carbohydrates occur largely in combination with proteins; mucoproteins and glycoproteins occur widely in mucosal structures.

Hyaluronic acid and chondroitin sulfuric acid are classified as mucopolysaccharides. In connective tissues they are bound to collagen and other proteins to form complex aggregates. Hyaline cartilage contains large quantities of sodium and potassium. For that reason, the sodium content of cartilage is higher than that of any other connective tissues (280 mEq/kg tissue water, or about twice the sodium concentration of blood plasma). Hyaluronic acid is soluble, and occurs in a free state in synovial fluid and certain other body fluids. The other components of the fluids are serum albumins and globulins. When synovial fluid is treated with acetic acid at pH 4, hyaluronic acid forms an insoluble mucin with the serum proteins. The isoelectric points of the proteins are in the neighborhood of pH 5; at a lower pH they are positively charged, and form the insoluble aggregate with the negatively charged mucopolysaccharide. The product is not a true glycoprotein, but a complex coacervate formed through salt linkages. It is readily redissolved by increasing the pH or the ionic strength.

Colloidal Aggregates

Complex aggregates are formed by two or more colloidal substances that are held together in a stable state of aggregation. They may include a water-rich phase, containing the colloids in solution; the two phases are in equilibrium. The formation of the aggregate depends on the pH and the electrolyte composition of the solution. Electrolytes are distributed between the two phases, the distribution depending on electrostatic and thermodynamic conditions of equilibrium. The life slimes of the extracellular ground substance and the intracellular cytoplasm, nucleus and stroma have been considered by many investigators to be heterogeneous aggregates of this nature. It has often been assumed that electrolyte equilibrium in cells and tissues is determined by the properties of these aggregates.

In a book published in 1906, Loeb expressed his point of view as follows: 'In 1899 I outlined a general theory of irritability which may be briefly summarized in the following sentences which I quote from a former paper: "The salts or electrolytes do not enter into this combination as a whole but through their ions. The great importance of the ion-proteid compounds (or soaps) lies in the fact that, by the substitution of one ion for another, the physical properties of the proteid compounds change (e.g., their surface tension, their power to absorb water, or their viscosity or state of matter). We thus possess in these ion-protein or soap compounds essential constituents of living matter, which can be modified at desire, and hence enable us to vary and control the life phenomena themselves. Life phenomena, and especially irritability depend on the presence in the tissues of a number of metal proteids, or soaps (Na, Ca, K and Mg) in definite proportions."'

LOEB describes many phenomena of irritability in muscle and other contractile tissues in relation to the ionic content of experimental solutions. In many cases there are definite antagonisms between the various cations: sodium eliciting muscular contractions, which are then inhibited by calcium, and so on. These phenomena are exhibited not only by muscle and other contractile tissues, but also by many other types of biological material. For example, the development of *fundulus* eggs is dependent on the electrolyte balance of the surrounding solution. LOEB's work on this problem represents a continuation of the studies undertaken by RINGER, and by HOFMEISTER. As shown by the foregoing quotation, LOEB considered the electrolyte balance within the cells and tissues to be conditioned by the external conditions such as electrolyte concentrations and pH.

HOFMEISTER's point of view was much the same, except that at the period during which he worked, the concepts of hydrogen ion concentration and pH had not been developed. It should be pointed out that at that time, the internal properties of the cellular matrix were considered to depend on the properties of the external environment. This was BERNARD's view, as expressed in his principle of the *milieu intérieur* of the higher vertebrates (chapter II). At present many investigators support a doctrine at variance with this, namely that the electrolyte content of the intracellular substance is maintained by energy derived from respiratory metabolism (active transport). The relation between electrolyte equilibrium or distribution and intracellular metabolism is a problem that involves both the statics and dynamics of living matter.

In the present study, electrolyte composition and distribution are regarded as primary functions and respiratory metabolism is regarded as secondary. As explained in chapter I,

$$A, B, C, T \supset M$$

To understand the properties of colloidal aggregates as they are exhibited by non-living systems *in vitro,* it is necessary to consider two aspects of the problem: (1) the thermodynamic properties, and (2) the electrostatic properties. The aggregates are mixtures of different colloids, water and electrolytes. Typically, they are formed by two electrically charged colloids of different sign. A nucleohistone would be a biologically important example. It is formed under conditions in which the histone is positively charged, forming an insoluble aggregate with the negatively charged nucleic acid. The supernatant solution contains the colloidal components in equilibrium with the insoluble aggregates. The colloid-rich phase contains water and the same neutral salts which are present in the supernatant solution. According to thermodynamic principles, two immiscible phases are formed under conditions in which the free energy of mixing becomes positive. This depends on the enthalpy of mixing and on configurational entropy.

The thermodynamic functions are related to the electrostatic properties and to other intermolecular forces of attraction through which the components interact. Since the system is heterogeneous, the distribution of electrolytes between the two phases is non-ideal, and characterized by imperfect mixing. That the water distribution is also non-ideal is shown by the fact that the water content of the colloid-rich phase is much lower than that of the supernatant solution. This implies a deviation from the laws of perfect mixing.

The chemical potential of water is the same in the two phases; the mole fraction in the colloid-rich phase is different. Therefore, there is a deviation of the chemical potential in the colloid-rich phase. In general, this is true of cells and tissues. Their water content is often in the vicinity of 70%. The water content of blood plasma and of other body fluids is greater than 90%. It is generally true of all kinds of phase equilibrium that the composition of any two phases may differ greatly. The exceptional cases of similar composition occur only in the neighborhood of transition points. As will be shown in chapter V, the total electrolyte concentration of intracellular phases is usually less than that of the *milieu intérieur*. This is due to the well-ordered state of the intracellular water, which produces a state of high configurational free energy and low configurational entropy. In systems in which water exists everywhere in the liquid state of aggregation, as in the classical Donnan equilibrium, there is generally an excess of inorganic ions in the protein solution, as compared with an equilibrium dialysate. Equilibrium then is characterized by a positive 'Donnan correction' in the osmotic pressure [LOEB, 1924; HENDERSON, 1928].

The manner in which proteins and other organic macromolecules form stable colloidal solutions depends on their hydrophilic properties. The protein molecule contains a number of hydrophilic groups (NH, OH, O), which have affinities for either the hydrogen or the oxygen atom of the water dipoles. These groups, as well as the ionized amino and carboxylate groups, combine with water-soluble substances and with water itself. The tendency of a hydrophilic colloid or 'sol' to combine with the solvent leads to the formation of a shell or 'membrane' of water dipoles enclosing the protein molecule. Any process which leads to dehydration of the hydrophilic macromolecules tends to coagulate the protein. When they lose the hydration shell, the molecules become free to form large aggregates; they then condense to form insoluble flocculent precipitates, which are stabilized by electrostatic and chemical attractions.

Mixtures of proteins in solution may be highly stable. It may be very difficult or impossible to separate the components. However, they may all be precipitated simultaneously by a process of coagulation in which all the components form a stable insoluble aggregate. The properties of such mixtures in either phase differ greatly from those of the individual components. SORENSEN [1930] has shown that in the isolation and purification of proteins, several fractions may be present, which show different compositions and solubilities. Mixtures of proteins or of other colloidal substances generally show properties which cannot be inferred from those of the isolated components. This may be explained by short-range interactions of the components resulting from electrostatic and chemical attractions involving all the kinds of macromolecules and the solvent. The resulting deviations from the solution laws would depend on configurational entropies. The problem is thus related to the general thermodynamics of mixtures. WILLSTATTER [quoted by OPARIN, 1938] emphasizes the importance of mixtures. 'Until now it was taken for granted in chemistry that the properties of components in chemical compounds disappear but that in mixtures of substances they are retained. This is an antiquated viewpoint... Mixtures may actually possess the nature of new chemical compounds.' OPARIN, in considering such properties of protoplasm as heat coagulation, surface precipitation, permeability and electrical properties, concludes that the effects are 'the resultant of correlation and reciprocal action of different colloidal systems which make up the protoplasm. This alone compels us, in considering the evolution of organic substance, to rely not upon those alterations in which this or another isolated compound may be subjected, but to bear in mind alterations which take place in complex mixtures of various organic substances.'

In addition to coagulation, there are other important types of separation of hydrophilic colloids from water. In the process known as *Entmischung* (unmixing or 'unscrambling'), the colloidal solution separates into two equilibrium layers, one of which is free of colloid. The phenomenon has been extensively studied by DE JONG, and has been designated *coacervation* (1949). It is characteristic that the colloid-rich 'coacervate' droplets are separated from the equilibrium fluid by a visible thin film of liquid. Droplets may fuse together, but they do not merge into the water phase. This illustrates the important principle that thin films of water represent states that differ from the bulk of the fluid.

The explanation of simple coacervation given by DE JONG is based on a change of the hydration layer of the colloidal particles in the hydrosol. The innermost layers of the shell are firmly bound to the colloidal particle, and are strongly oriented. At progressively greater distances there is more and more disorientation up to the limit where the water molecules are arranged like those of the pure solvent. Thus the boundary between the colloidal particle and the solvent is diffuse. In simple coacervates the boundary becomes sharp and limited. The particle is surrounded by a thin, well-defined shell. A number of such partially dehydrated particles may then condense to form coacervate particles which separate from the equilibrium liquid. In this special type of concentrated sol there is rigid orientation of the water dipoles, which become separated from the free molecules of the liquid. It is characteristic of certain kinds of protoplasm or slimes that, although they are relatively rich in water, they are immiscible with water or dilute salt solutions. This is explained by the fact that the shell of water is firmly bound and in a different state of aggregation from that of the free liquid.

In the complex coacervates formed by oppositely charged colloids, the behavior is determined by two antagonistic forces, the electrostatic attractions and the hydration shells. Complex coacervates are formed when the electrostatic attractions are strong enough to overcome the protective effect of the aqueous shell or membrane. When these two opposing forces are of nearly equal magnitudes, the system may be extremely labile. In that case, small changes of condition would affect the equilibrium. Thus minute changes of temperature, pH or electrolyte concentration tend to reverse the process of complex coacervation and to restore the shells of hydration that are characteristic of the hydrosols.

Stability of Biological Species

Biological time is measured on a scale of far greater magnitude than that of our familiar historical time scale. We have become familiar with geological time only within the last century. Some of DARWIN's antagonists dated all events from about 4,000 BC, a chronology based on the Book of Genesis. Today we may accept that date as a rough estimate of the earliest historical records, but we know that human prehistory extends from a more remote past – perhaps 100,000 or 200,000 years. Human evolution has been extremely rapid. Compared with the history of the social insects, such as the ants and termites, the time has been very brief, especially if it is measured in generations. To obtain a true idea of the stability of the gene, it is necessary to consider the geological rather than the human historical time scale.

Among the more stable species are the ants and termites. According to WHEELER [1923], ants which were embedded in the Baltic amber date from a period of 50 or 60 million years ago. When compared with living species, they are 'practically indistinguishable from those now living in Europe and North America'.

Among living species, there are even more ancient forms. Possibly the oldest is *Lingula,* a brachiopod, which is the oldest known animal genus. '*Lingula* still lives in sand bars and flats of the Chesapeake Bay under conditions which have not effected any essential change in its structure since the lower Cambrian (500,000,000 years). Who can look at a living *Lingula* without being overwhelmed by the effort to grasp its immeasurable antiquity?' [BROOKS, 1899]. Paleozoic scorpions (about 300,000,000 years) look much like some recent species, and *Limulus,* the king crab, has changed little since its appearance in the Triassic, about 200,000,000 years ago. Fossil crinoids (sea-lilies) were known long before any living forms had been seen, and were supposed to have vanished by the end of the Paleozoic. Living forms have been recovered, and although most are stemless and free swimming, there is an early stalked stage.

'The entire scientific world was amazed at the discovery in recent years of living specimens of a coelocanth fish, *Latimera chalumace,* thought to have become extinct at the end of the cretaceous era (50,000,000 years), looking remarkably like the reconstructions of its predecessors which got started in the Devonian (300,000,000 years). Some of its ancient relatives, the *Rhipiestia,* took to the land in the middle or late Devonian, but the coelocanths in their long history have never given rise to any other group' [MORROW, 1961].

These examples are perhaps sufficient to establish upper limits for the stability of the gene. However, the higher vertebrates evolve far more rapidly, and show many significant changes during the geological record.

Visitors to the Natural History Museum are familiar with the evolution of the horse from the early three-toed 'eohippus' to the contemporary variation. In his book 'The Next Million Years', DARWIN [1953] takes that period as a measure of man's historical future. 'It is hardly necessary to say that there is nothing very exact about this million years. Some species change more quickly and some more slowly, but it does seem to be a good enough rule, and curiously enough it seems to apply more or less irrespective of the number of generations in a million years, which of course would be immensely more for an insect or a rat than for a buffalo or a man.'

We may accept the opinion that biological time is on a vast but finite scale, and that the gene, although not immortal, is among the most stable of chemical substances. The number of genes is continually increasing; they duplicate each other and reproduce exact copies in the cellular nuclei. The function of the nucleus, to a large extent, is reproductive. The cells divide and multiply, reproducing cytoplasm and extracellular substances. The functions of the extracellular substances are largely nutritive and protective. They supply the cell with nutrients which enable it to grow, multiply, and convert energy and chemical substances. They also protect the cell from the rigors of environmental change by maintaining constant conditions within the organism. The cell nucleus is the best protected structure in the organism. It contains the chromosomes and genes, so essential to the life of the cell and the tissue. It is the gene which determines the future of the species. In protecting the gene by all available means, nature works to preserve and advance the species.

Stability of the Gene[1]

The rate of mutations is important in the tempo of evolution. It is possible to deal with this question on the basis of statistical theory and quantum theory as related to the rates of chemical change. Genes may be treated as molecules, and the latter may be treated as crystalline solids. In kinetic theory, the average energy of a gas atom at temperature T is 3/2 kT, where k is the Boltzmann constant. The stability of a molecule is dependent on temperature because of this relationship. Let us consider the probability at any temperature that a molecule may acquire sufficient energy W to 'lift' it out of its characteristic and most probable state at a

1 SCHRÖDINGER [1944].

lower temperature. By raising the temperature, not only the number of collisions with other molecules increases, but also the force of each impact. The probability that a molecule may acquire the energy W is porportional to the factor

$$\exp\left(\frac{-W}{kT}\right).$$

It follows that the probability increases exponentially with the absolute temperature. From this relation, an expression may be derived for the 'time of expectation' of chemical change. Denoting the time as t, then,

$$t = \alpha \exp\left(\frac{W}{kT}\right) \tag{3.1}$$

where α is a constant of the magnitude of 10^{-13} to 10^{-14} sec. The fact that molecules are stable depends on the condition that W is of a much higher order of magnitude than kT. If it were of the same order of magnitude, then the lifetime of the molecule would be of the order of 10^{-13} sec, an extreme limit of instability. The constant α has a physical significance. It corresponds to the order of magnitude of the period of vibrations occurring at all times in the molecular system.

From these elementary statistical considerations, it is possible to make some inferences regarding the stability of molecules. The 'expectation time', t, may be influenced by exposing the system to ordinary visible light, to X-rays, or to γ-radiations. Chemical change may also be brought about by high energy electrons or other particles. By these means, the chance that any molecule may acquire the energy W is greatly increased. The energy

Table V. Energy levels of radiation from infrared to γ-rays

Type of radiation	Range of wavelength	Wave number[1], cm^{-1}	Energy, kcal	Energy, eV
Infrared	30 μm	333	1.951	0.0412
	0.8 μm (8,000 Å)	12,500	35.8	1.55
Visible	8,000 Å	12,500	35.8	1.55
	4,000 Å	25,000	71.6	3.10
Ultraviolet	4,000 Å	25,000	71.6	3.10
	1,500 Å	66,600	190	8.25
X-rays and	50 to	2×10^6	5,720	247.8
γ-rays	0.001 Å	10^{11}	2.86×10^8	123.9×10^5

1 Wave number is reciprocal of wavelength.

may be expressed in calories, in electron volts or in joules, and these quantities may be expressed in relation to the frequency or to the wave length of the radiations (table V). It is evident that 1 eV corresponds to 23,060 cal/mol. This is equivalent to the energy of light quanta in the infra-red. In the UV spectrum (4,000–1,500 Å), the energy corresponds to values between 71.5 and 190 kcal/mol. Very hard X-rays (50 Å) impart energy equivalent to 5.72×10^3 kcal/mol. γ-rays are 50,000 times as strong. At 25 °C, the value of RT is approximately 600 cal. Now,

$$\frac{\text{Molal energy}}{\text{RT}} = \frac{\text{W}}{\text{kT}} \tag{3.2}$$

For hard X-rays,

$$\frac{5.72 \times 10^6}{600} = 9{,}530$$

When a light quantum or photon of this energy is absorbed by a solid, an amount of energy of the order of 10,000 times that of the kinetic energy of an atom is released. For genetic mutations this exceeds the order of magnitude of the factor W.

To obtain an idea of the relation between W and the stability of a molecule or gene, values of the expectation time, t, are given in relation to $\frac{\text{W}}{\text{kT}}$.

$\frac{\text{W}}{\text{kT}}$	t	W, eV
30	0.1 sec	0.9
50	16 months	1.5
60	30,000 years	1.8

According to SCHRÖDINGER [1944], after DELBRUCK.

Thus a change of configuration somewhere in a gene could be produced by a chance fluctuation of vibrational energy resulting from an impact equivalent to that of a 2-volt electron. The isomer is sufficiently stable to withstand the impact of 1.0- or 1.5-volt electrons. It is vulnerable to high energy electrons or to ionizing radiations in the UV or X-ray spectrum.

These considerations regarding mutation rates and the effects of radiation or electron impacts show that the gene is an extremely stable configuration of molecules. It must be regarded as an aperiodic crystal of very low entropy. Like all such crystals, it can reproduce exact replicas of itself. In the organism, it is to be regarded as a center of low entropy. This implies that it is also a center of order in the cell, and that it resists all the perturbations that produce disorder. By ordering the internal structure of the cell, it maintains form, function, and stability. Intracellular stability also depends on all the factors which tend to hold the extracellular structures and fluids in constant states. One of these factors is the constancy of the *milieu intérieur,* which is necessary in stabilizing the electrolyte composition of protoplasm.

Protoplasmic Electrolytes

Defining his concept of *phase* in a physicochemical system, GIBBS [1928] wrote: 'We may call such bodies as differ in composition or state different *phases* of the matter considered ...' In accordance with this definition, we may distinguish extracellular from intracellular phases, treating the *milieu intérieur* as extracellular and protoplasm as intracellular. Protoplasm and the *milieu intérieur* are made up of the same set of inorganic electrolyte components (sodium, potassium, calcium, magnesium and chloride). These are derived from the *milieu extérieur* of the primitive ancestral waters of paleozoic ages, as described in chapter II [MACALLUM, 1910, 1926]. The intracellular system is also characterized by a set of *n* characteristic macromolecular substances and water. These have no free or independent existence, but form colloidal mixtures or aggregates which behave as units characteristic of each kind of cell or tissue. In accordance with the ideas of WILLSTATTER and OPARIN described in the foregoing, each component of such a mixture loses its characteristic properties found in the free state. The macromolecular components of protoplasm, forming complex aggregates that also contain water and electrolytes, develop complex, coherent physicochemical and physiological properties that cannot be found in simpler aggregates or in simple models.

In the growth and development of living cells and tissues, the primary properties are chemical morphology and state, as defined by the $(n + 6)$ substances that are necessary to describe chemical composition [JOSEPH, 1971a]. These include the set of *n* macromolecules: the number 6 denotes water and the five kinds of ions. During the growth period, the entire set of $(n + 6)$ substances has only one degree of freedom. This means that at any stage of development, there is only one normal state of maximal stability or accessibility (chapter I). Biological development and aging are attended by changes in chemical morphology and state. Formation of the *n* colloidal components is well-ordered, taking place with only one degree of freedom in the growth period [JOSEPH, 1971a]. After maturity, protoplasm in any kind of cell may become invariant. The *milieu intérieur* of mammalian species tends to be an ontogenetic and phylogenetic invariant, differing in composition and state from protoplasm and the colloidal aggregates of connective tissue. The development of cells and tissues thus is not free and unconstrained, but is subject to definite thermodynamic and genetic conditions of invariance and constraint (chapter I).

One of the conditions of constraint in the physicochemical system is the electrolyte composition of sea water. This applies to organisms, the cells of which are directly exposed to the external environment, as is the case with the simplest metazoa. In higher forms such as fish and other vertebrates, the cells are in contact with internal body fluids or with blood or hemolymph. With the development of the vertebrate kidney, these fluids were brought under homeostatic control. The electrolyte composition described in chapter II is characterized by the ion concentrations $(c_{Na})'$, $(c_K)'$, $(c_{Ca})'$, $(c_{Ma})'$ and $(c_{Cl})'$. Typical values for skeletal muscle are given in table VI for the following species: man, the Roman eel and the lobster. These represent the following types of muscle: mammalian, marine vertebrate and invertebrate. Expressed with reference to 100 mol sodium, the values are:

	Na	K	Ca	Mg	Cl
Man	100	487	5.2	44.5	77
Eel	100	661	37.2	29.6	94.8
Lobster	100	200	6.2	22.9	132

As the most striking feature, these figures show the high tendency of muscle tissue to accumulate potassium ions at the expense of sodium. Thus the ratio of intracellular potassium to sodium may vary from 2.0 as in the lobster to 6.6, as in the Roman eel. The other cations, calcium and magnesium, show lower intramuscular concentrations than sodium. Magnesium varies from a ratio of 22.9 to one of 44.5 mol/100 mol sodium. Each kind of tissue is well-ordered with respect to ion concentrations, but there is greater variability in the intramuscular concentrations than is found in blood plasma (chapter II). This does not seem surprising when it is considered that the three species are so greatly different.

In the following chapters, electrolyte composition and distribution are treated as primary properties in the description of chemical morphology and state. Ion distribution is regarded as a measure of configurational entropy and free energy of protoplasm. Viewed in this way, it is also a measure of irritability, mechanical work and tension. These properties, like the dielectric constant and electrical potentials, are determined by the chemical morphology and state of protoplasm.

Chapter IV: Reversible and Irreversible Electromotive Force

In electrochemical theory, there is a difference of electrical potential at any heterogeneous boundary where there exists a difference of chemical potential of ions. Two kinds of boundary potentials can be distinguished – reversible and irreversible. In biological systems, a steady state of constant ion concentrations in two contiguous phases requires thermodynamic conditions of constraint at the phase boundary. Then the rate of transport of sodium, potassium, calcium, magnesium and chloride ions must be zero. Therefore, the work of transport and the free energy change must also be zero. These are the necessary conditions for homeostasis and constant chemical composition. The electromotive force at such a boundary is reversible. This is a condition for an invariant steady state.

Unlike the inorganic physiological ions, carbon dioxide and carbonic acid are products of cellular respiration and metabolism. In the most stable state they are produced from carbohydrates and fats at a constant rate that corresponds to the rate of evolution of heat, as expressed in calories per unit time. These are exothermic reactions that proceed spontaneously with a decrease of free energy. Conditions for homeostasis and invariant chemical morphology require spontaneous transport of oxygen, carbon dioxide and heat at highly coordinated rates, which are determined by the rates of the intracellular oxidation reactions. Bicarbonate and hydrogen ions therefore diffuse spontaneously from the cell under conditions that constrain sodium, potassium and chloride ions to reversible transport. It is a condition for energy and electrolyte balance that the electromotive force at the boundary be invariant, and that all processes of ion diffusion and exchange must be constrained by phase rule conditions of invariance [JOSEPH, 1971a, b]. This requires the reversible kind of electromotive force between any two invariant phases.

Chemical Potentials and Electrical Potentials

In general, reversible electromotive force between two phases in an invariant heterogeneous system is described by equations 687 and 688, as developed by GIBBS [1928]. This requires a balance between a change of chemical potential, $\Delta\mu_i$, for a given kind of ion, and the reversible electromotive force, E. Before stating the two equations for reversible electromotive force, it will be instructive to work out an analogous equation for equilibrium of oxygen and nitrogen in the gravitational field of the Earth's atmosphere. Equilibrium of either gas in the atmosphere depends on a balance of two forces: (1) the tendency of the gas to escape from the pressure of 1 atmosphere at sea level to a low pressure (i.e. $\frac{1}{2}$ atmosphere) at an altitude h, and (2) the gravitational force. The molal free energy of expansion from 1 to $\frac{1}{2}$ atmosphere is:

$$\Delta G_1 = RT \ln \tfrac{1}{2} \tag{4.1}$$

The work of transporting one mole of gas to a height h in the atmosphere is:

$$\Delta G_2 = Mgh \tag{4.2}$$

where M is the molecular weight and g is the acceleration of gravity. The condition for a steady state in the atmosphere at constant temperature T is:

$$\Delta G_1 + \Delta G_2 = 0 \tag{4.3a}$$

and

$$\ln p_h = -\frac{Mgh}{RT} \tag{4.3b}$$

where p_h is the pressure at height h. The reversible work of expansion in the isothermal system is zero (CARNOT's principle). Thus we may write:

Driving force + constraining force = 0

and state this as a general principle for reversible processes in invariant systems at constant temperature. In an invariant heterogeneous system, the driving force for any ion distributed between two phases is $\Delta\mu_i$ or $(\mu_{i''} - \mu_{i'})$. This tends to force the ion in the direction of the lower chemical potential, i.e. with a negative value of $\Delta\mu_i$. The constraining force at the boundary is represented by FE, where F is the Faraday electrochemical constant, 96,500 joules or 23,060 cal/Equiv. Therefore in a system containing the five physiological ions, there are five equations of balance or constrain:

$$FE + \Delta\mu_{Na} = 0$$
$$FE + \Delta\mu_{k} = 0$$
$$FE + \tfrac{1}{2}\Delta\mu_{Ca} = 0$$

$$FE + \tfrac{1}{2}\Delta\mu_{Mg} = 0$$
$$FE - \Delta\mu_{Cl} = 0 \qquad (4.4)$$

where both E and $\Delta\mu$ are reversible and invariant in a homeostatic steady state. Each equation states the condition for invariance, which requires that the reversible work of transport must be zero. By eliminating FE from the set of five equations, we obtain:

$$\Delta\mu_{Na} = \Delta\mu_K = \tfrac{1}{2}\Delta\mu_{Ca} = \tfrac{1}{2}\Delta\mu_{Mg} = -\Delta\mu_{Cl} = \delta \qquad (4.5)$$

where δ denotes the equivalent change of chemical potential of each kind of ion. It is given by the ratio of $\Delta\mu_i$ to z_i, where the charge z_i is taken as negative for anions. The four conditions that correlate the five values of $\Delta\mu_i$ indicate that the system may be either univariant or invariant. When it is invariant, both E and δ are constants that are independent of time. In a system that changes with time, as in growth, development and aging, there is only one degree of freedom. In that case, δ and E are univariant functions of the chemical morphology. In any case, the five values of $\Delta\mu_i$ are highly coordinated and controlled by phase rule conditions of constraint.

Irreversible Transport of Ions in Invariant Systems

The normal partial pressure of oxygen gas at the surface of the Earth is about $\frac{1}{5}$ atmosphere, or 150 mm Hg. If large quantities of the gas were released at a high pressure (10 atm) at the Earth's surface, it would eventually escape into the upper atmosphere by a process that involves a diminution of free energy. Thus:

$$\Delta G_1 + \Delta G_2 \leqslant 0 \qquad (4.6a)$$

and

$$RT \ln p_h + Mgh \leqslant 0 \qquad (4.6b)$$

Thus the driving force, $-\Delta G_1$, is no longer balanced by the constraining force, Mgh, and the gas escapes irreversibly into the upper atmosphere.

The continuous intracellular production of CO_2 and H_2CO_3, by an analogous process leads to irreversible escape to blood plasma, lungs and external environment, with a diminution of free energy. Therefore, the conditions of irreversible transport require that:

$$FE + \Delta\mu_H \geqslant 0; \quad FE - \Delta\mu_{HCO_3} \leqslant 0 \qquad (4.7)$$

where $\Delta\mu_H$ represents the difference $(\mu_{H''} - \mu_{H'})$, and $\Delta\mu_{HCO_3}$ denotes the difference $(\mu''_{HCO_3} - \mu'_{HCO_3})$.
Therefore:

$$\Delta\mu_H \geqslant \Delta\mu_{Na}; \quad \Delta\mu_{HCO_3} \geqslant \Delta\mu_{Cl} \tag{4.8}$$

Hence bicarbonate and hydrogen ions are produced continually and irreversibly from active cells and transferred to the *milieu intérieur* under conditions that constrain sodium, potassium and chloride ions to reversible transport. Conditions of invariance and constraint require carbonic acid to be conducted to the environment by circulation and pulmonary ventilation at the same rate at which it is produced. It is also required that the metabolic heat be conducted irreversibly to the environment at the same rate at which it is formed. The rates of heat production and conduction are exactly correlated with the rates of carbon dioxide formation and of oxygen consumption, as measured in calories per unit time.

These are the necessary conditions for physiological homeostasis, physicochemical invariance, and perfect calorimetry in the mammalian organism [JOSEPH, 1971a, b]. Perfect calorimetry in an invariant system is implied by the first law of thermodynamics – this connotes a state of energy and electrolyte balance on a day-to-day basis, under which fluctuations of state may occur during the 24-hour interval. Quantitative calorimetry is also implied by the second law of thermodynamics (CARNOT's principle), which requires that no internal work be produced in an isothermal invariant system. This principle would apply to the organism on a day-to-day basis, during which the *net* internal work must be zero, when taken between two states in which the organism is found to remain in the same standard state, or most accessible state.

Calculation of Thermodynamic Functions from Electromotive Force

The standard resting electromotive force, E, between an intracellular and an extracellular phase such as the *milieu intérieur* depends on the change of chemical potential, $\Delta\mu_i$, between the two phases (equation 4.4). The change of chemical or electrical potential between any two such phases depends only on the physicochemical state of each phase, and is independent of the conducting pathways between the two phases. Thus it is a secondary property, which depends only on the primary property, which is chemical morphology. It is independent of other secondary properties such as permeability or repiratory metabolism. This is true of all reversible thermodynamic functions in an invariant system; the fundamental principle is that for any reversible system $\Sigma\Delta\mu_i$ is zero over *any* closed pathway. This conforms to the requirement of Carnot's principle, according to which reversible isothermal work in an invariant system must be zero [JOSEPH, 1966a]. Therefore, in the following, the resting electromotive force, E, will be called simply 'resting electrical potential' or 'electromotive force' rather than by the term 'membrane potential'. The latter term has been current in biology for many years, and is derived from the classical Donnan equilibrium, which is strictly applicable only to true solutions in systems with perfect mixing. In heterogeneous biological systems the change of chemical potential of sodium between an intracellular phase and blood plasma is given by:

$$\Delta\mu_{Na} = \Delta\mu_{Na}^{\circ} + RT \ln \frac{c_{Na}''}{c_{Na}'} \quad (4.9)$$

In the standard resting state of an invariant biological system, the activity coefficients of sodium ion, f_{Na}'' and f_{Na}', may be assigned unit values [JOSEPH, *et al.*, 1961, 1964, 1965; JOSEPH, 1966a, b, 1971a, b]. In ordinary salt solutions the value of 1.0 is assigned to the activity coefficient, f_i, of any ion at infinite dilution of all electrolytes. This rule cannot be applied to biological systems, in which the concept of a reference state at 'infinite dilution' is meaningless. In the living organism in its standard state, all ionic concentrations are nearly invariant in each kind of cell or tissue. Hence the reference state can only be the normal standard resting state of the whole organism [JOSEPH, 1971a, b]. In that state f_i for any ion is assigned the value of 1.0. Then the deviations from the classical Donnan equilibrium or from the laws of perfect mixing are expressed by the values of $\Delta\mu_i^{\circ}$ for any ion. Then, according to equations 4.4 and 4.9, for sodium ion:

$$-FE = \Delta\mu_{Na}^{\circ} + RT \ln \frac{c_{Na}''}{c_{Na}'} \quad (4.10)$$

Therefore $\Delta\mu_{Na}^{\circ}$ can be estimated from a known value of the resting reversible electromotive force, E, when the values of c_{Na}'' and c_{Na}' are known [JOSEPH, 1966b, 1971a, b]. From equation 2.4 and the known values of z_i and b_i (table IV), it is possible to estimate the values of $\Delta\mu_i^{\circ}$ and the intracellular values of D'' (table VII). Calculations have been made for the five ions in skeletal muscle of young and adult rabbit, adult rat and cat, and for the muscle of the marine crab, *Carcinus maenas*. The ratio c_{Na}''/c_{Na}'

is expressed as r_{Na}, as obtained from analytical values in the literature. The calculations are based on equation 4.10 and on the Born equation 2.4.

The results for adult mammalian skeletal muscle show values for $\Delta\mu_{Na}^{\circ}$ of 2.7 or 2.8 kcal. A value of 2.7 kcal is likewise found for the skeletal muscle of *Carcinus maenas*. The values of E for mammalian muscle are −80 mV, as compared with −58 mV for carcinus muscle. The invertebrate muscle shows a value of 0.10 for r_{Na}, as compared with values of 0.20–0.26 found in mammalian muscle. The dielectric constant D″ is 29 or 30 for both mammalian and invertebrate muscle. The values of all these quantities for the newborn rabbit are quite different. The sodium distribution ratio, r_{Na}, is 0.76, as compared to values of 0.20 for the adult rabbit and 0.10 for the marine crab. Dielectric constant is also higher, 44 as compared with 29 for the adult rabbit. For $\Delta\mu_{Na}^{\circ}$ the values are 1.3 kcal for newborn rabbits and 2.8 kcal for adults. Values of the standard chemical potentials for the other ions are highly correlated with that of $\Delta\mu_{Na}^{\circ}$ in each kind of muscle. The electrolyte composition of vertebrates and invertebrates is also characterized by well-ordered sets of the five kinds of ions (table VI). In both phylogenetic and ontogenetic development, the five values of $\Delta\mu_i^{\circ}$ are correlated by equations 4.4 and 4.5. In either kind of development, the values of the electrical potentials are also univariant. They are correlated with the five univariant values of $\Delta\mu_i$ [Joseph, 1971a, b].

The Donnan Equilibrium

Equations 4.4 and 4.5 represent the general conditions of constraint in *any* heterogeneous system that is either invariant or univariant. The system is invariant when δ and E are constants. It is univariant when there is a one to one relationship between δ, E and time. Thus it is univariant

Table VI. Electrolyte composition of skeletal muscle (mM/kg water

Species	Na	K	Ca	Mg	Cl
Man[1]	28.7	134.6	1.5	12.8	22.0
Roman eel[2]	25.0	165.2	9.3	7.4	23.7
Lobster[3]	83.2	166.6	5.2	19.1	109.9

1 Dickerson and Widdowson [1960].
2 Robertson [1960].
3 Robertson [1961].

in normal growth, development and aging, but may approximate a state of invariance when the rates of chemical and morphological change are very slow, as they are after the end of the normal growth period in mammals [JOSEPH, 1971a, b].

In any given case, the electromotive force E and the distribution of the inorganic electrolytes depend on the values of $\Delta\mu_{Na}^{\circ}$ and $\Delta\mu_i^{\circ}$ for the other four ions. For all ions, these values depend on the dielectric constant, D″. Then, according to equation 2.4 for any ion:

$$\Delta\mu_i^{\circ} = \frac{164\, z_i^2}{b_i}\left(\frac{1}{D''} - \frac{1}{80}\right) \tag{4.11}$$

where $\Delta\mu_i^{\circ}$ represents the molal work (W_i''–W_i') of transferring the given kind of ion from a solvent of dielectric constant 80 (water or any homogeneous body fluid) to a structured phase of dielectric constant D″. Thus $\Delta\mu_i^{\circ}$ is a measure of configurational free energy in the system. As D″ approaches 80 in any kind of phase, the five values of $\Delta\mu_i^{\circ}$ approach zero. This is characteristic of the conditions for the Donnan equilibrium [DONNAN, 1911]. In this example of the classical 'membrane equilibrium', each phase is a homogeneous aqueous solution characterized by a dielectric constant of 80. The heterogeneous character of the system is determined by the nature of the semi-permeable membrane, which must be structured and insoluble. Otherwise the two phases are perfectly miscible. The Donnan equilibrium thus applies only to systems to which a uni-state theory of water structure is applicable. Such systems, aside from the insoluble membrane, are characterized by the property of perfect mixing.

The general theory of GIBBS [1928], on the other hand, applies to *any* heterogeneous system, in which water in the various phases can coexist in any possible states of aggregation. Such systems are characterized by the intrinsic thermodynamic properties of all components and all phases. The distribution of all components is subject to general phase rule conditions of invariance and constraint, as well as to special conditions characteristic of any given system. In systems in which $\Delta\mu_i^{\circ}$ approaches zero for each kind of ion, equations 4.9 and 4.10 reduce to:

$$\Delta\mu_i = RT \ln \frac{c_i''}{c_i'} \tag{4.12}$$

and

$$-FE = RT \ln \frac{c_i''}{c_i'} \tag{4.13}$$

Equation 4.13 may be referred to as the reversible Nernst equation for a system in which $\Delta\mu_i^\circ$ is zero. It is not to be confused with the general equation 4.4. The equilibrium as described by equations 4.12 and 4.13 is reversible and conforms to the condition that:

Driving force + constraining force = 0

where the constraining force, FE, is measured either in joules or calories. Then the general condition of invariance or constraint leads to the general law of ionic distribution in the Donnan equilibrium:

$$r_{Na} = r_K = r_{Ca}^{\frac{1}{2}} = r_{Mg}^{\frac{1}{2}} = r_{Cl}^{-1} \quad (4.14)$$

where r_i for any ion denotes c_i''/c_i'.

This represents the perfectly non-selective distribution of ions in a two-phases system that is characterized by the perfect mixing of ions in each homogeneous phase. In such a system, the dielectric constant of each aqueous phase is approximately 80, and $\Delta\mu_i^\circ$ for each ion approaches zero. Therefore, changes of state of water in growth, development and aging cannot be accounted for in such systems. The results of table VII, for example, could not be explained on the basis of equations 4.12 and 4.13, in which the terms for configurational free energy and entropy are omitted. Changes of state in physiological responses are likewise outside the conceptual scheme that is embraced by the Donnan equilibrium.

The reason for this is that the scheme does not acknowledge the existence of configurational free energy and entropy that are necessary to account for the properties of all heterogeneous systems that do not show the property of perfect mixing. Such systems in living organisms are generally of a solid or semi-solid state of aggregation [SCHRODINGER, 1944; JOSEPH, 1971a, b]. Structured phases cannot be described by the conditions necessary for the classical Donnan equilibrium. In structured systems it is necessary to introduce $\Delta\mu_i^\circ$ for each kind of ion as in equations 4.9 and 4.10. These values depend on the apparent dielectric constant of the structured water, and are measures of configurational free energy and entropy in the system. As is shown in table VII, values of D″ differ for newborn and adult states in the rat, but are nearly constant for adult mammalian skeletal muscle. A similar value is found for the marine crab, *C. maenas.* The presence of two ordered states of water in muscle or other cellular phases permits intracellular changes of state. These can give rise to physiological responses such as contraction and action potentials and to changes of tension and production of external work. The system then

Table VII. Calculation of standard chemical potentials and dielectric constant from electromotive force

EMF dielectric constant and change of standard chemical potential	Young rabbit[1]	Adult rabbit[1]	Rat[1]	Cat[1]	*Carcinus maenas*[2]
E, mV	–48	–80	–80	–80	–59
r_{Na}	0.76	0.20	0.21	0.26	0.10
$\Delta\mu_{Na}^{\circ}$, kcal	1.3	2.8	2.8	2.7	2.7
$\Delta\mu_{K}^{\circ}$	0.9	2.1	2.1	2.1	2.1
$\Delta\mu_{Ca}^{\circ}$	4.5	10.7	10.7	10.0	10.0
$\Delta\mu_{Mg}^{\circ}$	6.5	14.3	14.3	13.8	13.8
$\Delta\mu_{Cl}^{\circ}$	0.7	1.5	1.5	1.5	1.5
D″	44	29	29	30	30

1 Chemical composition. WIDDOWSON and DICKERSON [1964]; EMF, FUDEL-OSIPOVA and MARTYNENKO [1964].

2 SHAW [1966, 1958a, b].

behaves as a unit, with a change of state that is characterized by one degree of freedom. The primary properties of the system may be denoted, as before, as A, B, C . . ., which refer to morphology and state. Action potentials, tension and work are then included in the secondary set, X, Y, Z ... In an all or nothing response such as is characteristic of skeletal muscle, the contracted state represents a normally accessible state characterized by a modified set of secondary properties, X′, Y′, Z′ ...

Action Potentials and Configurational Free Energy

The foregoing theory was developed in a study of the aging of mammalian skeletal muscle [JOSEPH, 1971a]. There it was shown that the following equations could serve as the basis for the observed resting, action and spike potentials. The following terms are expressed in electrical units: $\Delta\mu_{Na}$, $\Delta\mu_{Na}^{\circ}$ and RT ln r_{Na}, where r_{Na} is the ion distribution ratio, c_{Na}''/c_{Na}' (equation 4.9 and 4.10). Then:

$$-FE = -FE^{\circ} - RT \ln r_{Na} \quad (4.15a)$$

$$FE^{\circ} + \Delta\mu_{Na}^{\circ} = 0 \quad (4.15b)$$

$$FE_{Na} + RT \ln r_{Na} = 0 \quad (4.15c)$$

E° and E_{Na} are defined by the equations. The spike potential E_{Na} is proportional to $- RT \ln r_{Na}$, and the action potential E_a is equal to $- E^\circ$.

The values of E, E_a and E° are conventionally expressed in millivolts.
The values may be converted to kcal by the conversion factor 1 kcal = 43.4 mV, or 100 mV = 2.306 kcal. Typical values are of the following orders of magnitude in mammalian skeletal muscle: $E = -80$ mV; $E_a = 120$ mV; $E_{Na} = 40$ mV. The corresponding values in kcal are:

$\Delta\mu = 1.84$; $\Delta\mu_{Na}^\circ = 2.7$ or 2.8 kcal, as estimated in table VII. These values were calculated from the resting potentials for the muscle of rat, rabbit and cat, using observed values of r_{Na}.

According to this theory, the action potential E_a is a direct measure of $\Delta\mu_{Na}^\circ$. Accordingly, it is also a direct measure of configurational free energy or entropy of the stretched fiber in its resting state. Contraction involves an increase of configurational entropy due to a simultaneous change of water, contractile proteins, and electrolytes. The system shows one degree of freedom, with simultaneous changes of electrical, mechanical and chemical properties. The change of state of the water would also lead to changes of permeability of the cell wall or sarcolemma and in all other states of water.
transport properties which depend on the intracellular and membranous

In a system with one degree of freedom, all these properties would depend on the physicochemical state of the entire system. They would be included among a set that describes a large number of secondary properties, X, Y, Z ..., all of which change simultaneously with muscular stimulation and contraction. Any one of these properties, X, for example, would change to a stimulated value, X′. The change (X′ − X) is then a measure of irritability or responsiveness.

A set of simultaneous physiological responses can readily be explained on the basis of a bi- or multi-state theory of intracellular water. This is the problem that was posed historically by STENSEN [1662] and by VON HALLER [1757] (chapter I). It cannot be explained on the basis of a theory in which water cannot change state, remaining always characterized by a dielectric constant of 80. The latter state is that of a homogeneous, unstructured aqueous solution. The uni-state theory (Donnan equilibrium) is incompatible with the views of modern histology, according to which intracellular water is part of a highly organized and structured system [FREY-WYSSLING, 1953].

Irreversible Electrical Potentials

In addition to the reversible kinds of electrical potentials represented by equations 4.4 and 4.10, there exists the possibility of irreversible types. These would occur under conditions in which the four stabilizing conditions given by equations 4.5 and 4.14 do not apply. The following expression is applicable, in general, both to reversible and irreversible differences of potential at phase boundaries [GLASSTONE, 1960; JOSEPH, 1971a, b].

$$FE + \Sigma \frac{t_i}{z_i} \Delta\mu_i = 0 \tag{4.16}$$

where t_i is the transport number of the *i*th kind of ion. This represents the fraction of current carried by the given kind of conducting ion. Summed over all ions, $\Sigma t_i = 1.0$. In an invariant system $\Delta\mu_i/z_i$ is constant for all ions; therefore in the limiting case:

$$FE + \Delta\mu_i/z_i = 0 \tag{4.17}$$

This is equivalent to equation 4.4. Under these conditions, E is constant and reversible, and the system is in a state of homeostasis. When equations 4.4 and 4.17 are not applicable, the system may acquire as many as five degrees of freedom. The values of $\Delta\mu_i/z_i$ become independent, and ion diffusion at the boundary becomes free and unconstrained. Under such conditions, E and $\Delta\mu_i$ become irreversible, and E is described by equation 4.16.

In such a system there are five independent values of $\Delta\mu_i$ and five independent values of t_i. The value of E thus depends on at least five independent variables. The values of $\Delta\mu_i$ are correlated with five values of $\Delta\mu_i^\circ$ and five values of c_i''. In a system of such complexity, there would be no general kind of solution of equation 4.16. The only possible general solution requires a one-to-one relation between E and $\Delta\mu_i$. This can be found only when E and $\Delta\mu_i$ are reversible, as in equations 4.4 and 4.17. These conditions apply to the stable resting potentials in invariant systems with phase rule conditions of constraint that apply to the reversible diffusion of all ions that are not produced in respiratory metabolism. Under the same conditions of constraint, the irreversible diffusion of bicarbonate and hydrogen ions is governed by equation 4.7.

Equation 4.17 yields the classical Donnan equilibrium and the classical Nernst equation only under the limiting condition that $\Delta\mu_i^\circ$ becomes zero for all the ions (equation 4.12–4.14). This requires a liquid state of aggregation of water in both phases and a uniform dielectric constant of about

80. Such systems are never to be confused with actual heterogeneous systems where the water exists in two or more states of aggregation that are determined by the thermodynamic properties of two or more immiscible phases. Then the water may exist or coexist in solid or semi-solid states of aggregation, characterized by dielectric constants that are much lower than 80. In such bi- or multi-states of aggregation, there are considerable supplies of configurational free energy, which may be released in changes of state of water, electrolytes and polyelectrolytes. Such changes of state would involve mechanical work and tension, changes of electrical potential, and changes of dielectric properties. Such properties are also involved in growth, aging and in development of chemical morphology. In these changes, chemical morphology and state are described by the set of primary properties A, B, C ... Resting and action potentials are members of the set of secondary properties which depend on chemical morphology and state. It is important never to confuse the primary and secondary properties. It is equally important always to distinguish systems in which water may exist in two or more states from systems in which water is homogeneous and in which the electrolytes are dissolved in solutions with perfect mixing.

Chapter V: Colligative Properties

The colligative properties of water in homogeneous aqueous solutions are those that can be directly derived from the chemical potential of water. Four properties of this kind are usually considered to be: freezing point depression, boiling point elevation, vapor pressure lowering, and osmotic pressure. Through thermodynamic equations, any one of these properties may be derived from any of the other three. Thus each secondary property is a function of the primary property, which is the composition and state of a given solution. Applying the same principle to a vertebrate or invertebrate organism, it could be said that the colligative properties of the *milieu intérieur*, the blood, hemolymph and body fluids depend on the standard state that is characteristic of the organism under standard conditions of composition and state.

Thus mammalian blood is characterized by a freezing point depression of about 0.56–0.60 °C. These values correspond to a lowering of the chemical potential of water by about 3.0–3.16 cal/mol. Constant blood composition thus implies constant freezing points and constant chemical potentials of water in the mammalian organism. According to the phase rule, this fixes the standard value of the chemical potential of water at about −3.1 cal/mol in mammalian cells and tissues [Joseph, 1971a, b]. Osmotic pressure is related to freezing point by a direct one to one thermodynamic relationship.

In a heterogeneous system in which the state of water is constant throughout, but in which protein and electrolyte composition is determined by the conditions of the classical Donnan equilibrium, the condition for a constant chemical potential of water is maintained by the osmotic pressure of each phase. This condition corresponds to van't Hoff's law of osmotic pressure, as applied for example by Sorensen [1914] and by Loeb [1924].

In heterogeneous systems in which water may coexist in more than one state of aggregation at different values of the dielectric constant, equilibrium may occur under conditions of constant temperature and pressure throughout the system. In this type of system, there are variations of configurational entropy and free energy in any structured phase. These correspond to variations of composition and state of electrolytes and water, as they are determined by the nature and state of the *n* macromolecular substances that are characteristic of each phase. Each phase is characterized by a standard state of configurational free energy and entropy, and by a standard content of water and electrolytes.

When a given phase is in the liquid state, it is a solution characterized by perfect mixing. In this case, the configurational free energy is zero, when referred to any other liquid state. Then equilibrium is maintained by osmotic pressure, and the system responds only by changes of pressure. However, when a phase is of a solid or semi-solid character, water may coexist in two or more states and the response may be one that involves only the configurational thermodynamic functions such as entropy or free energy, with no change of pressure or volume. These responses would imply changes of dielectric constant and changes of chemical and electrical potentials of the ions. They would thus lead to the production of action potentials and to transfer and redistribution of energy and entropy in the system, depending on the exact conditions of constraint. Thus the change of physicochemical state of water in cells or tissues leads to univariant responses of the type found in physiological systems that include nerve, cardiac or skeletal muscle.

From what has been said, it follows that all the secondary physicochemical and physiological properties of cells and tissues depend on the chemical morphology, which is the primary property. The first function of any cell or tissue is to exist in its characteristic standard state. This depends on a curve of growth, development and aging. As was pointed out in chapter I, this embryological and ontogenetic development occurs under two kinds of constraints: genetic and thermodynamic. Physicochemical constraints are imposed by the phase rule, which requires constant chemical potentials of water and electrolytes in all cells and tissues, including the *milieu intérieur* in its standard state. A state of water balance at any time requires a corresponding state of electrolyte balance, of configurational free energy and entropy, and of an indefinitely extended set of secondary properties.

Liquid Systems

In a two-phase system, which includes a semi-permeable membrane, water may be in the liquid state in each of the two phases. In such a system, the membrane is in a solid state of aggregation, and the state of water may differ from that in either of the liquid phases. Even in such a limiting case, the water may not constitute a uniform or homogeneous dielectric. Nevertheless, the colligative properties of the two solutions may be considered to be those of homogeneous aqueous solutions. If the system contains electrolytes in each phase, and a non-dialyzable polyelectrolyte or protein in one phase, the equilibrium can be treated along the lines established by DONNAN [1911]. The following treatment, however, follows GIBBS [1928] rather than DONNAN. According to GIBBS' equation 97 (the Gibbs-Duhem equation):

$$-S'dT + V'dP = m'_{H_2O}d\mu'_{H_2O} + \Sigma m_i'd\mu_i' \tag{5.1}$$

In the outer solution, or equilibrium dialysate, $m_{i'}$ refers to mass of any ion expressed as moles, and $\mu_{i'}$ denotes the corresponding chemical potential. The mass of the water in moles is m'_{H_2O}, and $\mu_{H_2O}'$is the chemical potential of water in the dialysate. Entropy is expressed by S′, volume by V′, pressure by P, and absolute temperature by T.

A similar equation applies to phase 2, the inner solution that contains the soluble protein.

$$-S''dT + V''dP = m_{H_2O}''d\mu_{H_2O} + \Sigma m_i''d\mu_i'' + m_2 d\mu_2'' \tag{5.2}$$

where m_2'' and μ_2'' refer to the mass and chemical potential of the soluble protein. In the absence of any electrolyte, $\Sigma m_i''$ and $\Sigma m_i'$ become zero. Then at constant temperature and constant $_{H_2O'}$

$$d\mu'_{H_2O} = d\mu''_{H_2O} = 0. \tag{5.3}$$

This condition represents a constant value of μ_{H_2O} in the two-phase system (GIBBS' equation 78). Accordingly:

$$V''\,dP'' = m_c\,d\mu_c \tag{5.4}$$

For any undissociated solute:

$$d\mu_c = RT\,d\ln m_c \tag{5.5}$$

Therefore

$$dP'' = \frac{RT}{V''}\,d\,m_c \tag{5.6}$$

Equation 5.6 is van't Hoff's law of osmotic pressure, as applied to a salt-free protein solution. For the outer dialysate containing a total of $\Sigma m_i'$ moles of free ions:

$$dP' = RT\, \Sigma dc_i' \tag{5.7}$$

where the concentration is expressed as moles per liter.

The contribution of the ions to the osmotic pressure was called the 'Donnan correction' by LOEB [1924]. It is here expressed as:

$$P_{corr.} = RT\, [\Sigma c_i'' - \Sigma c_i'] \tag{5.8}$$

The theoretical osmotic pressure is then given by:

$$P = RT\, c_2 \;\; + RT[\Sigma c_i'' - \Sigma c_i'] \tag{5.9a}$$

or

$$P = \bar{P} + P_{corr.} \tag{5.9b}$$

In the Donnan equilibrium, $P_{corr.}$ is always positive. As HENDERSON [1928] showed, for any binary electrolyte such as HCl or NaCl

$$c_H'' + c_{Cl}'' \geqslant c_H' + c_{Cl}' \tag{5.10a}$$

or

$$c_{Na}'' + c_{Cl}'' \geqslant c_{Na}' + c_{Cl}' \tag{5.10b}$$

This follows from the mathematical fact that the square is the rectangle of smallest perimeter. In phase 1,

$$c_{Na}' + c_H' = c_{Cl}'$$

In an acid solution of phase 2, where the protein is positively charged:

$$c_{Na}'' + c_H'' \geqslant c_{Cl}''$$

The inequalities of equation 5.10 follow from the above relations when it is noted that:

$$r_H = r_{Na} = r^{-1}{}_{Cl}$$

as has been shown in equation 4.14. Whereas in the classical Donnan equilibrium, the term $\Sigma(c_i'' - c_i')$ is always positive, in most intracellular phases, it is negative [JOSEPH, 1971a, b]. The term is a measure of configurational free energy in the two-phase system; this is quite high in cellular phases in which the water is in well-ordered structured forms of low entropy and low dielectric constant. Then the standard chemical potentials of the ions are high, leading to low intracellular concentrations of sodium and chloride ions and to high concentrations of potassium in the bound form. The resting electrical potentials are then highly negative (about −80 mV), and in muscle, the action potentials are high (chapter IV).

By way of contrast, in cartilage and certain other kinds of connective tissue, the values of $\Sigma(c_i'' - c_i')$ may be highly positive. In these tissues, the water is not highly organized, the dielectric constants are high, resting electrical potentials may approach zero, and configurational free energy is low. Connective tissues, therefore, in certain cases may conform to the classical Donnan conditions, which require the existence of water in one state of aggregation, with a uniform dielectric constant. In other cases, the state of water, as measured by the dielectric constant, may approach that of certain cellular phases such as kidney or fetal skeletal muscle. This may be true of certain states of skin, in which the dielectric constant may be less than 70, which is also characteristic of kidney and fetal muscle. Thus there is no essential difference in the theoretical thermodynamic treatment of intracellular and extracellular phases.

Solid Aggregates

In solid states of aggregation, the dielectric constant D″ of the structured water may be much less than 80. In that case, the water is at the same chemical potential as that of a liquid phase with which the solid phase is in equilibrium. Then, according to GIBBS' equation 77:

$$\mu'_{H_2O} = \mu''_{H_2O}. \tag{5.11}$$

The value of μ'_{H_2O} in a liquid phase of the *milieu intérieur* is directly related to the freezing point depression, Δ. At constant pressure and composition μ'_{H_2O} depends also on the temperature. According to equation 5.1:

$$- S'dT = d\mu'_{H_2O} \tag{5.12}$$

for 1 mol of water. Therefore, in the equilibrium with ice:

$$\mu'_{H_2O} = \mu^\circ_{H_2O} + \Delta S_f dT_f \tag{5.13}$$

where ΔS_f is the entropy of fusion, T_f is the freezing point of the solution, μ'_{H_2O} is the chemical potential of water in the solution, and $\mu^\circ_{H_2O}$ is the chemical potential of pure water. Therefore:

$$\mu'_{H_2O} = \mu^\circ_{H_2O} + 5.26\,\Delta \tag{5.14}$$

where the entropy of fusion ΔS_f is 5.26 cal/mol/deg.

Let us imagine an aqueous electrolyte solution of the same total molal concentration $\Sigma(c_i)''$ as that of an intracellular phase. Its freezing point depression $\bar{\Delta}$ would differ from Δ because, in general $\Sigma(c_i)''$ is less than $\Sigma c_i'$. This is a result of the high configurational free energy of structured intracellular phases. By the same token, the equilibrium values of μ'_{H_2O} and μ''_{H_2O} would be lower than $\bar{\mu}_{H_2O}$, the chemical potential of water in the aqueous solution of total concentration $\Sigma(c_i)''$. It is evident from equation 5.14 that:

$$\mu'_{H_2O} - \mu_{H_2O} = 5.26\,(\Delta - \bar{\Delta}) \qquad (5.15)$$

from which it follows that the difference $(\mu'_{H_2O} - \mu_{H_2O})$ is negative. For a system at constant temperature and pressure, the Gibbs-Duhem equation 5.1 is written in the form:

$$-\,m_{H_2O}\,d\mu_{H_2O} = \Sigma m_i\,d\mu_i \qquad (5.16)$$

where μ_i is the chemical potential of the *i*th kind of ion, and m_i is the mass, expressed as moles. The number of moles of water is expressed as m_{H_2O}. Expressing the relationship with respect to ion concentrations in moles per kg water:

$$-\,d\mu_{H_2O} = 0.018\,\Sigma c_i\,d\mu_i \qquad (5.17)$$

where c_i is the concentration of the *i*th kind of ion, and 0.018 is the molecular weight of water divided by 1,000.

By integration of equation 5.16:

$$\mu_{H_2O}'' - \mu_{H_2O} = 0.018\,RT[\Sigma(c_i)'' - \Sigma c_i']. \qquad (5.18)$$

The differences of potential and concentration depend on the change of *standard chemical potentials* of the ions between the two phases. In an intracellular phase, the ionization constant α_i'' is defined in relation to the total concentration $(c_i)''$ by:

$$-\,d\mu''_{H_2O} = 0.018\,\Sigma\alpha_i''(c_i)''\,d\mu_i'' \qquad (5.19)$$

Therefore:

$$\bar{\mu}_{H_2O} - \mu''_{H_2O} = 0.018 \int \Sigma\alpha_i''(c_i)''\,d\mu_i'' \qquad (5.20)$$

In the change of state from the aqueous solution of total concentration $\Sigma(c_i)''$ to an intracellular system of the same concentration, there are simultaneous changes of dielectric constant and standard chemical potentials. For any ion, the differential $d\mu_i$ depends on a change of $\bar{\mu}$, the ideal part, and on a change of standard chemical potential $\Delta\mu_i^\circ$, which represents the configurational term. For any ion:

$$d\mu_i = d\bar{\mu}_i + d\mu_i^\circ \qquad (5.21)$$

and

$$d\mu_i = RT\, d\ln \alpha_i + RT\, d\ln (c_i)'' + d\mu_i^\circ \qquad (5.22)$$

In the integration $(c_i)''$ is taken as constant for each ion and for potassium and chloride the integrals

$$\int \alpha_K''(c_K)''\, d\mu_K'' \text{ and } \int \alpha_{Cl}''(c_{Cl})''\, d\mu_{Cl}''$$

are taken as zero. For the chloride ion, the approximation is made that

$$\Delta\mu_{Cl}^\circ + RT \ln \alpha_{Cl}'' = 0 \qquad (5.23)$$

For potassium, the value of α_K'' is small (of the order of 0.02–0.06) at all values of $\Delta\mu_K^\circ$. Applying these considerations, and taking the value of α_{Na}'' as 1.0:

$$c_{Na}''\Delta\mu_{Na}^\circ = RT\,[\Sigma c_i' - \Sigma(c_i)''] \qquad (5.24)$$

The product $c_{Na}''\Delta\mu_{Na}^\circ$ is a measure of the *dielectric energy* or *configurational free energy* of the system. In intracellular phases it is highly positive. A typical value for human skeletal muscle is of the order of 80 cal/kg muscle water.

At 37 °C the value of RT is 620 cal. The ionic concentration of plasma is about 0.315 mol/kg water, and the intracellular value of $\Sigma(c_i(''$ is about 0.185 mol. Hence:

$$c_{Na}''\ \Delta\mu_{Na}^\circ = 620 \times 0.130 = 80.6 \text{ cal/kg water} \qquad (5.25)$$

The value of c_{Na}'' is 28.7 mM/kg water, and this yields a value of 2.81 kcal for $\Delta\mu_{Na}^\circ$ [JOSEPH, 1971a]. The corresponding value of the dielectric constant D″ is 31. The calculated action potential, E_a, is 116.6 mV and the calculated resting potential E is −79.3 mV. The various values are estimated from c_{Na}'' and $\Delta\mu_{Na}^\circ$ by the use of equations 4.15a, b, c.

Primary and Secondary Properties

Temperature, pressure and chemical composition of any intracellular or extracellular phase may be taken as the primary properties which are necessary to describe chemical morphology and state. As was explained in chapter I, these constitute a set of (n + 8) properties; at constant temperature and pressure, the number is reduced to (n + 6). Thus warm-blooded species (mammals and birds) have two less degrees of freedom than cold-blooded species, such as deep sea fish and invertebrates, flourishing in regions of high pressure and low temperature.

In mammals, the five inorganic ions and water are characterized by five values of m_i and one value of m_{H_2O}. There are five corresponding values of the conjugate variables, μ_i, and one of μ_{H_2O}. Ion concentrations are expressed by the five ratios m_i to m_{H_2O}, and these are related to the five values of μ_i. The value of $\mu^\circ_{H_2O}$ in any mammalian species is a phylogenetic and ontogenetic invariant [JOSEPH, 1971a].

Thus there are five invariant chemical potentials in the system: one for water and four for the electrolyte components. The latter fall into two classes: binary and ternary. The binary electrolytes include NaCl and KCl, which may be described as type AB. The ternary electrolytes are of type AB_2 and include $CaCL_2$ and $MgCl_2$. For these electrolytes:

$$\mu_{AB} = \mu_A + \mu_B$$
$$\mu_{AB_2} = \mu_A + 2\mu_B \tag{5.26}$$

Therefore, if each potential μ_i is regarded as a function of chemical morphology, there are five values of μ_i for the ionic species: sodium, potassium, calcium, magnesium and chloride. These form a set of five secondary properties that determine the resting electrical potential E, according to equation 4.4. For each of the ions, there exists also a value of $\Delta\mu_i^\circ$ and a corresponding value of α_i''. These depend on the (n + 6) primary properties. In skeletal muscle and nerve, the action potential E_a is directly related to chemical morphology and state, and to the dielectric constant D''. Accordingly, the following quantities are to be included in the set of secondary properties: the resting potential, E, the action potential, E_a, the dielectric constant, D'', and the configurational free energy or dielectric energy.

All of these quantities depend on the intracellular state of water, as it is determined by the set of *n* macromolecular components. This set develops with the embryological change of colloidal composition that occurs in the fetal state and in the post-partum growth period. In these periods, growth and development is not free and unconstrained. When there is a well-ordered one-to-one correspondence between chemical morphology and chronological age, the system has only one degree of freedom. This corresponds to the fact that the synthesis of the *n* substances is well-ordered and synchronous.

The system develops as a unified and organized mixture in which each macromolecular substance loses its identity as a chemical individual. Accordingly, protoplasm develops as a unit, in which each of the component macromolecules tends to lose its own chemical idiosyncracies, as it contributes to the physicochemical properties of the entire macroassembly of protoplasm as a whole [WILLSTATTER, quoted by OPARIN, 1933]. The well-ordered synthesis of such an assembly of *n* macromolecular substances is not free and unconstrained, but occurs as the development of a set, having only one degree of freedom [JOSEPH, 1971a].

In a similar way, phylogenetic and ontogenetic development depends on the synchronized formation of sets of intracellular and extracellular macromolecular polyelectrolytes, rather than on the free and unconstrained development of individual molecules. This development is subject to phase rule conditions of invariance and constraint, which tend to maintain constant composition and state in the *milieu intérieur.* Constrained development of protoplasm requires two kinds of control – genetic and thermodynamic. Since intracellular water is well-ordered and structured, there are permissible changes of state which involve concomitant changes of configurational free energy or dielectric energy. These depend on states of negative entropy related to water structure.

A bi- or multi-state theory of intracellular water structure can explain the properties of cellular 'irritability' or 'responsiveness', which cannot be explained by the properties of water in a homogeneous system with perfect mixing.

The primary and secondary properties of cells and tissues require for their description the general principles of the 'equilibrium of heterogeneous systems', as worked out by GIBBS [1928]. They are incompatible with the limiting case that was worked out by DONNAN [1911]. The latter theory cannot account for irritability and responsiveness, which depend on a change of state of structured water. The explanation of these physicochemical and physiological behavioral responses requires a theory that expresses the existence of water in two or more labile states of aggregation.

It follows from what has been said that the liquid solution phases of heterogeneous systems are restricted rather than unrestricted phases or true solutions. Accordingly, the *milieu intérieur* of mammals, as described by BERNARD [1878] and by MACALLUM [1910, 1926], tend to be invariant phases. Therefore, they are not treated theoretically as the true solutions of physical chemistry, which are unrestricted systems, that may have any arbitrarily large number of degrees of freedom [JOSEPH, 1971a, b]. For unrestricted systems, the chemical potential of sodium μ_{Na}, is expressed by the relation:

$$\mu_{Na} = \mu_{Na}^{\circ} + RT \ln c_{Na} f_{Na} \qquad (5.27)$$

where f_{Na} is the activity coefficient. In ordinary salt solutions it is defined in relation to a reference state at infinite dilution, where the value of f_{Na} becomes 1.0. At any finite concentration, the value depends on the ionic strength of the solution; this is determined by the magnitude of the interionic forces [DEBYE and HUCKEL, 1923]. The value of $\Delta\mu_{Na}^{\circ}$ between any two phases in an invariant heterogeneous biological system depends mainly on the intracellular dielectric constant D″. In such systems the reference state for the definition of f_{Na} and all other ionic activity coefficients can only be the physiological state. In this state, all values of f_i may be assigned standard values of 1.0. All values of $\Delta\mu_i$ are then determined by equations of the form:

$$\Delta\mu_i = \Delta\mu_i^{\circ} + RT \ln \frac{c_i''}{c_i'} \qquad (5.28)$$

where $\Delta\mu_i^{\circ}$ depends on D″ and the magnitude of the interionic forces in the two phases. Generally, in intracellular phases, $\Delta\mu_i^{\circ}$ depends on the dielectric constant rather than on the interionic attractions, which yield terms of smaller orders of magnitude. However, in connective tissues, where D″ approaches the value of 80 characteristic of water, $\Delta\mu_i^{\circ}$ is determined by the interionic forces, especially in tissues such as cartilage, which are characterized by a highly negative colloidal charge. Further consideration of such interionic attractions will be given in later chapters under the topics of acid-base balance, colloidal charge and interionic forces. Thus in all cells and tissues, $\Delta\mu_i$ and other forms of configurational free energy and entropy depend on dielectric constant, acid-base balance and colloidal charge. Since each of these parameters depends only on chemical morphology, they are interdependent rather than independent functions. Biological systems therefore tend to be invariant or univariant rather than bivariant or polyvariant. They are subject to many thermodynamic conditions of constraint. In this respect, they differ from the true solutions of pure physical chemistry, which are unrestricted, and which may have many degrees of freedom.

Chapter VI: Skeletal Muscle

Of all cells and tissues, skeletal muscle is the one in which configurational free energy is maximal and most highly adapted toward conversion to external work. This implies a well-ordered state of water, in which the dielectric constant is low and the standard chemical potentials of the ions are high. High configurational free energy implies a low intrafibrillar concentration of sodium, highly negative resting potential, and an action potential of the order of 120 mV. It also connotes a value of $\Delta\mu_{Na}{}^{\circ}$ of the order of 3.0 kcal.

In the following pages, a number of these properties will be calculated from the water and electrolyte composition of vertebrate and invertebrate skeletal muscle. It will be shown not only that the sodium concentration is low, but also that the total intracellular electrolyte concentration $\Sigma(c_i)'$ is much lower than the value for blood plasma $\Sigma c_i'$. According to the theoretical methods developed in chapter V, it is then possible to estimate the chemical potentials, standard chemical potentials and dielectric properties of intracellular phases. These properties depend on the existence of intrafibrillar water in two or more immiscible states of aggregation. This idea agrees with those of the classical theory of STENSEN [1662], but is incompatible with a uni-state theory of water aggregation on which the 'membrane equilibrium' theory of DONNAN [1911] was based. It is also incompatible with the classical Nernst equation, as applied by DONNAN [1911] and by LOEB [1924]. The Nernst equation, as was shown in chapter V, meets the condition for electrochemical reversibility, but takes no account of the existence of intracellular water in two or more coexistent states of aggregation.

Valid equations for resting and action potentials must satisfy not only the conditions for thermodynamic and electrochemical reversibility, but must also take into consideration changes in state of intracellular water and electrolytes. The classical Donnan and Nernst equations for ion distributions and electrical potentials do not consider the configurational free energy and 'negative entropy' that are inherent in all well-ordered and structured biological systems. Therefore, they disregard changes of configurational entropy or free energy. Well-ordered and structured systems have the properties of irritability and responsiveness. By changes of structure or state, they are capable of responding to external or internal stimuli.

From GLISSON [1654], STENSEN [1662] and VON HALLER [1757] to HUXLEY [1868] and PEIRCE [1892], irritability and responsiveness to stimuli have been recognized as the unique properties of living matter or protoplasm. These properties are easily explained by labile intracellular states of structured water. They are outside the conceptual scheme embraced by DONNAN's theory of 'membrane equilibrium', which does not consider the variable dielectric properties of cellular phases. This limitation is not suffered by GIBBS' theory of heterogeneous equilibrium, which embraces all possible kinds of thermodynamic state and changes of state [GIBBS, 1928].

Table VIII. Free energies of ion distribution in human skeletal muscle[1]

	Na	K	Ca	Mg	Cl
Blood plasma, mEq	140	5	5	2.5	105
Tissue	28.7	134.6	3.1	25.6	22.0
Ratio (r_i)	0.20	26.9	0.62	10.2	0.21
RT ln (r_i), kcal	–0.99	2.03	–0.30	1.43	–0.96
$\Delta\mu_i^\circ$, kcal	2.81	2.10	10.5	14.0	1.54
$\Delta\mu_i$, kcal	1.82	1.82	3.64	3.64	–1.82
$(\Delta\mu_i^\circ)'$, kcal	2.81	–0.21	1.94	2.21	–0.86
$(\Delta G_{i,Na}^\circ)'$, kcal	–	–3.02	–0.84	–1.71	1.95
α_i''/α_i'	1.0	0.023	2.3×10^{-5}	5×10^{-9}	0.020

1 DICKERSON and WIDDOWSON [1960]; JOSEPH [1971a].

Table IX. Free energies of ion distribution in pig skeletal muscle[1]

	Na	K	Ca	Mg	Cl
Blood plasma, mEq	140	5	5	2.5	105
Tissue	26.8	136	2.8	29.1	17.6
Ratio	0.19	27.2	0.56	11.6	0.17
RT ln (r_i), kcal	–1.01	2.04	–0.36	1.51	–1.09
$\Delta\mu_i^\circ$, kcal	3.12	2.34	11.7	15.6	1.72
$\Delta\mu_i$, kcal	2.11	2.11	4.22	4.22	–2.11
$(\Delta\mu_i^\circ)'$, kcal	3.12	0.07	4.58	2.71	–1.02
$(\Delta G_{i,Na}^\circ)'$, kcal	–	–3.05	–0.83	–1.77	2.10
α_i''/α_i'	1.0	0.030	10^{-5}	8×10^{-10}	0.017

1 DICKERSON and WIDDOWSON [1960]; JOSEPH [1971a].

Mammalian Skeletal Muscle

Data for the electrolyte balance in four species of adult mammalian skeletal muscle are given in tables VIII–XI. The species are man, pig, rabbit and rat. For convenience the values taken for the ion concentrations in blood plasma are the same for all species. They represent the average values as given in table I, but the values are rounded out to two significant figures. The intramuscular values of the five ions (sodium, potassium, calcium, magnesium and chloride) are expressed as mEq/kg water. Total water is expressed as g/kg fresh tissue; the values, like those for the five ions, represent results for whole, fresh fat-free muscle. From the results for whole muscle and blood plasma, values for the distribution ratios (r_i) for the five

Table X. Free energies of ion distribution in rabbit skeletal muscle[1]

	Na	K	Ca	Mg	Cl
Blood plasma, mEq	140	5	5	2.5	105
Tissue	27.7	144.5	5.8	31.5	22.1
Ratio (r_i)	0.20	29	1.16	12.6	0.21
RT ln (r_i), kcal	–0.99	2.10	0.07	1.56	–0.96
$\Delta\mu_i^\circ$, kcal	2.68	2.01	10.0	13.4	1.47
$\Delta\mu_i$, kcal	1.69	1.69	3.38	3.38	–1.69
$(\Delta\mu_i^\circ)'$, kcal	2.68	–0.41	3.31	1.82	–0.73
$(\Delta G^\circ_{i,Na})'$, kcal	–	–3.09	–1.03	–1.77	1.95
α_i''/α_i'	1.0	0.020	1.6×10^{-5}	6×10^{-9}	0.028

1 WIDDOWSON and DICKERSON [1964]; JOSEPH [1971a].

Table XI. Free energies of ion distribution in rat muscle[1]

	Na	K	Ca	Mg	Cl
Blood plasma, mEq	140	5	5	2.5	105
Tissue	30.0	140.2	5.5	31.0	20.0
Ratio (r_i)	0.21	28	1.1	12.4	0.19
RT ln (r_i), kcal	–0.97	2.08	0.03	1.56	–1.02
$\Delta\mu_i^\circ$, kcal	2.58	1.93	9.6	12.9	1.42
$\Delta\mu_i$, kcal	1.61	1.61	3.22	3.22	–1.61
$(\Delta\mu_i^\circ)'$, kcal	2.58	–0.47	3.19	1.66	–0.59
$(\Delta G^\circ_{i,Na})'$, kcal	–	–3.05	–0.98	–1.75	2.03
α_i''/α_i'	1.0	0.021	3×10^{-5}	9×10^{-8}	0.039

1 WIDDOWSON and DICKERSON [1964]; JOSEPH [1971a].

ions are estimated. Each of these represents the total internal concentration $(c_i)''$ divided by the plasma concentration, $(c_i)'$. Each value represents the total of free plus bound ions in the given phase.

A fourth row of figures represents the value of $-$ RT ln (r_i) for each ion. At 38 °C this is calculated as $-$ 1.419 $\log_{10}(r_i)$, and the values are given as kcal. The value of $\Delta\mu_{Na}^{\circ}$, the change of standard chemical potential of sodium ion, is calculated from equation 5.25. This is written as:

$$\Delta\mu_{Na}^{\circ} = \frac{RT\,[\Sigma c_i' - \Sigma(c_i)'']}{c_{Na}''} \qquad (6.1;\,5.25)$$

The value of $\Sigma(c_i)''$ is calculated as $(Na^+ + K^+ + Cl^-)$, and the value of $\Sigma c_i'$ is taken as 0.315 mol/kg water. Then, according to the calculations of table VIII,

$$\Delta\mu_{Na}^{\circ} = \frac{620 \times (0.315\text{–}0.185)}{0.0287} = 2.81\ \text{kcal}$$

The corresponding values of $\Delta\mu_i^{\circ}$ for the other four ions are then obtained from the values of z_i^2/b_i, according to the method given in chapter II, according to which the values, as referred to sodium, are:

Cl	K	Na	Ca	Mg
0.55	0.75	1.0	3.75	5.0

The calculated values of $\Delta\mu_i^{\circ}$ for the other four ions, expressed as kcal, are entered in the fifth row of tables VIII–XI. The value of $\Delta\mu_{Na}$ is then calculated from the equation:

$$\Delta\mu_{Na} = \Delta\mu_{Na}^{\circ} + RT\ln(r_{Na}) \qquad (6.2)$$

Then:

$$\Delta\mu_{Na} = \Delta\mu_K = \tfrac{1}{2}\Delta\mu_{Ca} = \tfrac{1}{2}\Delta\mu_{Mg} = -\,\Delta\mu_{Cl} = \delta \qquad (6.3;\,4.5)$$

where δ is the equivalent change of chemical potential, expressed as kcal. Accordingly, $\Delta\mu$ for each of the five ions can be obtained from equations 6.2 and 6.3.

The values are given in the sixth row of tables VIII–XI. The value of $(\Delta\mu_i^{\circ})'$, the apparent change of standard chemical potential, is then derived from the relationships:

$$\Delta\mu_i = (\Delta\mu_i^{\circ})' + RT\ln(r_i) \qquad (6.4)$$

where $(\Delta\mu_i^{\circ})'$ is defined as:

$$(\Delta\mu_i^\circ)' = \Delta\mu_i^\circ + RT \ln \frac{\alpha_i''}{\alpha_i'} \tag{6.5}$$

and where α_i'' represents the ionized fraction $c_i''/(c_i)''$, and α_i' represents $c_i'/(c_i)'$. Therefore, according to equations 6.4 and 6.5:

$$\Delta\mu_i = \Delta\mu_i^\circ + RT \ln (r_i) + RT \ln \alpha_i'' \tag{6.6}$$

Accordingly, the values of α_i'' may be estimated from the other three values in equation 6.6, from the figures given in the fourth, fifth and sixth rows of tables VIII–XI. The value of α_{Na}'' is taken as 1.0 to indicate complete ionization of the intramuscular sodium [JOSEPH, 1971a, b]. This has been found to yield values of the resting and action potentials that agree with the results of experiment.

The value of the apparent standard free energy of ion distribution, as referred to sodium is expressed as:

$$(\Delta G^\circ{}_{i,Na})' = -RT \ln \frac{(r_i)^{1/z_i}}{r_{Na}} \tag{6.7}$$

$$(\Delta G^\circ{}_{i,Na})' = \frac{(\Delta\mu_i^\circ)'}{z_i} - \Delta\mu_{Na}^\circ \tag{6.8}$$

For NaCl, this yields:

$$(\Delta G^\circ{}_{Na,Cl})' = -RT \ln (r_{Na})(r_{Cl}) \tag{6.9}$$

For potassium-sodium distribution:

$$(\Delta G^\circ{}_{K,Na})' = -RT \ln \frac{(r_K)}{(r_{Na})} \tag{6.10a}$$

For calcium:

$$(\Delta G^\circ{}_{\frac{1}{2}Ca,Na})' = -RT \ln \frac{(r_{Ca})^{\frac{1}{2}}}{(r_{Na})} \tag{6.10b}$$

and for magnesium:

$$(\Delta G^\circ{}_{\frac{1}{2}Mg,Na})' = -RT \ln \frac{(r_{Mg})^{\frac{1}{2}}}{(r_{Na})} \tag{6.10c}$$

These four values are estimated in the eighth row of tables VIII–XI from the five values of $(\Delta\mu_i^\circ)'$, as given in the seventh row. Finally, the values of α_i'' for the various ions, as estimated from equation 6.6, are given in the tenth row. It is possible then to calculate the following secondary properties of the cell from the primary properties, as represented by the water and electrolyte content: the changes of standard chemical potentials, $\Delta\mu_i^\circ$, the apparent changes of standard chemical potential, $(\Delta\mu_i^\circ)'$, the change of chemical potential, $\Delta\mu_i$, for each of the five ions, and the apparent standard free energies of ion distribution. These secondary properties form a set of functions which can be inferred from the primary functions, as represented by the five distribution ratios, (r_i). From these functions, an indefinitely large number of secondary functions can also be inferred, as will be shown in subsequent chapters. These all stand in relation to the chemical morphology as predicates to subject. The nature of the intracellular protoplasm thus implies an indefinitely large number of secondary functions, X, Y, Z . . . Then:

$$A, B, C \ldots \supset X, Y, Z \ldots$$

where the primary functions represent the ion concentrations and distribution ratios. Secondary functions would include, in addition to those

Table XII. Free energies of ion distribution in frog muscle[1]

	Na	K	Ca	Mg	Cl
Blood plasma, mEq	103.8	2.5	3.0	2.0	74.3
Tissue	29.3	96.7	4.4	11.8	13.8
Ratio (r_i)	0.28	38.7	1.47	5.9	0.18
RT ln (r_i), kcal	0.73	2.19	0.24	1.06	1.00
$\Delta\mu_i°$, kcal	2.35	1.76	8.80	11.8	1.30
$\Delta\mu_i$, kcal	1.62	1.52	3.24	3.24	−1.62
$(\Delta G°_{i,Na})'$, kcal	2.35	−0.57	2.97	2.18	−0.62
$(\Delta G°_{i,Na})'$, kcal	–	−2.92	−0.89	−1.26	1.73
α_i''/α_i'	1.0	0.004	6×10^{-5}	8×10^{-8}	0.011

1 KATZ [1895]; FENN [1935].

Table XIII. Free energies of ion distribution in salmon muscle[1]

	Na	K	Cl
Blood plasma, mEq	211.9	3.60	156.7
Tissue	28.1	206	6.12
Ratio (r_i)	0.13	57.2	0.039
RT ln (r_i), kcal	−1.21	2.45	−1.92
$\Delta\mu_i°$, kcal	2.57	1.93	1.41
$\Delta\mu_i$, kcal	1.36	1.36	−1.36
$(\Delta\mu_i°)'$, kcal	2.57	−1.09	−0.56
$(\Delta G°_{i,Na})$, kcal	–	−3.12	2.01
α_i''/α_i'	1.0	0.007	0.037

1 PARRY [1961].

listed, electrical potentials, dielectric energy, configurational free energy, work and mechanical tension.

Cold-Blooded Vertebrates

In the calculation of the standard free energies of electrolyte distribution in the muscles of cold-blooded vertebrates the same formulations are applied as those used for mammals. In tables XII–XVI, the following kinds of muscle are described: skeletal muscle of the frog and salmon, tongue and parietal muscle of the eel, *Anguilla anguilla,* and skeletal muscle of the Roman eel. Values of $\Delta\mu_{Na}{}^{\circ}$ are estimated from the difference

Table XIV. Free energies of ion distribution in tongue muscle of the eel, *Anguilla anguilla*[1]

	Na	K	Ca	Mg	Cl
Blood plasma, mEq	183	3.2	2.4	3.8	155
Tissue	46.1	124.1	4.9	30.0	40.3
Ratio (r_i)	0.25	38.8	2.0	7.9	0.26
RT ln (r_i), kcal	–0.82	2.16	0.41	1.23	–0.79
$\Delta\mu_i{}^{\circ}$, kcal	2.16	1.62	8.1	10.8	1.19
$\Delta\mu_i$, kcal	1.34	1.34	2.68	2.68	–1.34
$(\Delta\mu_i{}^{\circ})'$, kcal	2.16	–0.82	2.27	1.45	–0.55
$(\Delta G^{\circ}{}_{i,Na})'$, kcal	–	–2.98	–1.03	–1.44	1.61
α_i''/α_i'	1.0	0.0036	5.4×10^{-5}	2×10^{-7}	0.059

1 CHAN *et al.* [1967].

Table XV. Free energies of ion distribution in parietal muscle of the eel, *Anguilla anguilla*[1]

	Na	K	Ca	Mg	Cl
Blood plasma, mEq	183	3.2	2.4	3.8	155
Tissue	40	152	11	91	25
Ratio (r_i)	0.22	47.5	4.6	28	0.16
RT ln (r_i), kcal	–0.89	2.30	–0.92	0.31	–1.11
$\Delta\mu_i{}^{\circ}$, kcal	2.05	1.53	7.65	–0.68	1.22
$\Delta\mu_i$, kcal	1.16	1.16	2.32	10.25	–1.16
$(\Delta\mu_i{}^{\circ})'$, kcal	2.05	–1.14	3.24	2.32	–0.05
$(\Delta G^{\circ}{}_{i,Na})'$, kcal	–	–3.19	–0.43	–0.89	2.00
α_i''/α_i'	1.0	0.010	7×10^{-4}	5×10^{-6}	0.10

1 CHAN *et al.* [1967].

Table XVI. Free energies of ion distribution in muscle of the Roman eel[1]

	Na	K	Ca	Mg	Cl
Blood plasma, mEq	212	1.95	3.87	2.2	188
Tissue	25.0	165.2	9.3	7.4	23.7
Ratio (r_i)	0.21	83	2.3	3.4	0.12
RT ln (r_i), kcal	–0.95	2.69	0.50	0.74	–1.92
$\Delta\mu_i^\circ$, kcal	2.60	1.95	9.8	13.0	1.43
$\Delta\mu_i$, kcal	1.65	1.65	3.30	3.30	–1.65
$(\Delta\mu_i^\circ)'$, kcal	2.60	–1.04	2.80	2.56	0.27
$(\Delta G^\circ_{i,Na})'$, kcal	–	–3.64	–1.20	–1.32	2.87
α_i''/α_i'	1.0	0.009	7×10^{-6}	7×10^{-9}	0.158

1 ROBERTSON [1960].

$[\Sigma c_i' - \Sigma(c_i)'']$ from equation 6.1, as was done for mammalian skeletal muscle. The values of blood sodium range from 103.8 mEq/l in the frog to 212 mEq/l in the Roman eel; the other ion concentrations are roughly proportional to sodium concentrations. The total concentrations $\Sigma c_i'$ range from 255 mM (frog), as estimated from the freezing point depression, to 408 mM in the Roman eel.

Calculated values of $\Delta\mu_{Na}^\circ$ range from 2.05 kcal in parietal muscle of Anguilla to 2.60 kcal in skeletal muscle of the Roman eel. Thus the values are somewhat lower than the corresponding values for mammalian skeletal muscle, as given in tables VIII–XI, where $\Delta\mu_{Na}^\circ$ is of the order of 2.8 kcal. The values of $(\Delta G^\circ_{K,Na})'$, the apparent standard free energies of potassium-sodium distribution, range from -2.92 (frog) to -3.64 kcal (Roman eel). The corresponding limits for mammalian muscle are -2.59 (rat) to -3.12 kcal (pig). Thus, on the whole, skeletal muscle of the cold-blooded species evince a somewhat greater tendency to accumulate potassium than that of the mammals. With the exception of the frog, and tongue of Anguilla, the intramuscular potassium is higher than that of mammalian muscle. However, in all cold-blooded species, the plasma potassium is *lower* than that of mammalian plasma. It is this factor that is responsible for the highly negative values of $(\Delta G^\circ_{K,Na})'$. It is also responsible for the highly negative values of $(\Delta\mu_K^\circ)'$. In mammals, the values of $(\Delta\mu_K^\circ)'$ range from 0.07 (pig) to -0.47 kcal (rat). In poikilotherms the corresponding limits are -0.55 (salmon) to -1.14 kcal in parietal muscle of the eel.

The differences are due largely to the greater tendency of poikilotherm muscle to bind potassium. Thus for all species except salmon, the values of α_K'' are of the order of 0.01 or less. The value for salmon is 0.007. The values of α_K'' in mammalian muscle range from 0.020 (rabbit) to 0.030 (pig). Values of α_{Ca}'', α_{Mg}'' and α_{Cl}'' have also been calculated for homoiotherms and will be compared in a later section.

Invertebrate Muscle

Data for electrolyte balance in several kinds of invertebrate muscle are given in tables XVII–XXI, inclusive. These include the marine crab, *Carcinus maenas,* the mollusc, *Mytilus edulis,* the scallop, *Pecten,* and the lobster, *Nephrops norvegicus.* In these species, the total ion concentrations of blood plasma are much higher than in the vertebrate species, where

Table XVII. Free energies of ion distribution in the muscle of the marine crab, *Carcinus maenas*[1]

	Na	K	Ca	Mg	Cl
Blood plasma, mEq	468	12.1	17.5	23.5	524
Tissue	54	120	6.9	17.9	54
Ratio (r_i)	0.11	10.0	0.40	0.76	0.10
RT ln (r_i), kcal	–1.32	1.37	–0.55	–0.16	–1.37
$\Delta\mu_i^\circ$, kcal	2.68	2.00	10.0	13.4	1.47
$\Delta\mu_i$, kcal	1.36	1.36	2.72	2.72	–1.36
$(\Delta\mu_i^\circ)'$, kcal	2.68	–0.01	3.27	2.88	0.01
$(\Delta G^\circ_{i,Na})'$, kcal	–	–2.69	–1.05	–1.24	2.69
α_i''/α_i'	1.0	0.035	8×10^{-5}	2×10^{-8}	0.085

1 SHAW [1955, 1958a, b].

Table XVIII. Free energies of ion distribution in muscle of pecten[1]

	Na	K	Cl
Blood plasma, mEq	490	12.5	573
Tissue	79	151	66
Ratio (r_i)	0.16	12.6	0.087
RT ln (r_i), kcal	–1.10	1.51	–1.40
$\Delta\mu_i^\circ$, kcal	2.60	1.95	1.43
$\Delta\mu_i$, kcal	1.50	1.50	–1.50
$(\Delta\mu_i^\circ)'$, kcal	2.60	–0.01	0.10
$(\Delta G^\circ_{i,Na})'$, kcal	–	–2.61	2.70
α_i''/α_i'	1.0	0.038	0.073

1 POTTS [1958].

Table XIX. Free energies of ion distribution in muscle of *mytilus edulis*[1]

	Na	K	Cl
Blood plasma, mEq	502	12.7	525
Tissue	54	120	54
Ratio (r_i)	0.11	9.4	0.092
RT ln (r_i), kcal	−1.30	1.32	−1.43
$\Delta\mu_i^\circ$, kcal	2.60	1.95	1.43
$\Delta\mu_i$, kcal	1.30	1.30	−1.30
$(\Delta\mu_i^\circ)'$, kcal	2.60	−0.02	0.13
$(\Delta G^\circ_{itNa})'$, kcal	–	−2.62	2.73
α_i''/α_i'	1.0	0.036	0.102

1 Potts [1955].

Table XX. Free energies of ion distribution in whole muscle of lobster *(Nephrops norvegicus)*[1]

	Na	K	Ca	Mg	Cl
Blood plasma, mM	517	8.6	16.2	10.4	527
Tissue, mM	83.2	166.6	5.2	19.1	109.9
Ratio (r_i)	0.16	19.3	0.32	1.85	0.21
RT ln (r_i), kcal	−1.07	1.75	−0.71	0.40	−0.96
$\Delta\mu_i^\circ$, kcal	2.60	1.95	9.98	13.0	1.43
$\Delta\mu_i$, kcal	1.53	1.53	3.06	3.06	−1.53
$(\Delta\mu_i^\circ)'$, kcal	2.60	−0.22	3.77	2.66	−0.57
$(\Delta G^\circ_{i,Na})'$, kcal	–	−2.82	−0.71	−1.26	2.03
α_i''/α_i'	1.0	0.026	2.5×10^{-5}	3×10^{-8}	0.035

1 Robertson [1961].

the freezing points are of the order of −0.6 to −0.8 °C, corresponding to values of $\Sigma c_i'$ that are of the order of 0.35–0.41 molal. In invertebrate blood the values of $\Sigma c_i'$ are 1 molal or higher, and the freezing depressions are correspondingly high. Consequently, the chemical potentials of water are quite low, of the order of −8.5 cal/mol as referred to pure water. It follows that the intracellular values, μ_{H_2O}'', are also low; balance is maintained in the intracellular phases by the presence of several kinds of nitrogen compounds such as amino acids, amines, and arginine phosphate, which are present in solution, and which have an osmoregulatory function [Robertson, 1961; Joseph, 1968].

In species with high extracellular electrolyte concentrations, the calcu-

Table XXI. Free energies of ion distribution in intracellular phases of lobster muscle *(Nephrops norvegicus)*[1]

	Na	K	Ca	Mg	Cl
Blood plasma, mM	517	8.6	16.2	10.4	527
Tissue, mM	24.5	188	3.72	20.3	55.1
Ratio (r_i)	0.047	21.9	0.23	1.97	0.10
RT ln (r_i), kcal	–1.82	1.83	–0.88	0.41	–1.37
$\Delta\mu_i°$, kcal	2.80	2.10	10.5	14.0	1.54
$\Delta\mu_i$, kcal	0.98	0.98	1.96	1.96	–0.98
$(\Delta\mu_i°)'$, kcal	2.80	–0.85	2.84	1.55	0.39
$(\Delta G°_{i,Na}$, kcal	–	–3.65	–1.38	–1.61	3.19
α_i''/α_i'	1.0	0.006	2.5×10^{-6}	10^{-9}	0.14

1 ROBERTSON [1961].

lation of $\Delta\mu_{Na}°$ cannot be based on equation 6.1, which is valid only for tissues in which no organic nitrogeneous osmoregulation is required. In such tissues equation 6.1 would require corrections for the concentrations, dissociation constants and standard chemical potentials of all such intracellular solutes. However, it is possible in such cases to estimate $\Delta\mu_{Na}°$ from a known value of the electrical potential E and a known value of the sodium distribution ratio, r_{Na}. Thus:

$$- FE = \Delta\mu_{Na}° + RT \ln r_{Na} \tag{6.11}$$

For example, the resting potential in Carcinus muscle is −59 mV, and the value of r_{Na} is 0.11. This yields a calculated value of 2.69 kcal for $\Delta\mu_{Na}°$ in carcinus muscle, as compared with values of the same order of magnitude in mammalian muscle (tables VIII–XI). The mammalian values range from 2.58 (rat) to 3.12 kcal (pig), as estimated from the electrolyte concentrations by means of equation 6.1. It is logical then to estimate the value of $\Delta\mu_{Na}°$ for the whole muscle of *mytilus, pecten,* and *N. norvegicus.* A value of 2.80 kcal, however, is assumed for the intracellular phase of lobster muscle. This is consonant with the results that are given in table XXI. When the values of the apparent standard free energies of potassium-sodium distribution $(\Delta G°_{K,Na})'$ are estimated, very similar results are obtained for Carcinus, Mytilus and Pecten muscle. These range from −2.61 for Pecten to −2.68 kcal for Carcinus. These may be compared with the values −2.82 and −3.65 kcal for lobster, the constants referring respectively to whole muscle and to the intracellular phases. It seems evident from these results that the total muscle water exists in at least two or more different states.

This agrees with the results of studies of NMR spectroscopy on mammalian muscle by HAZLEWOOD *et al.* [1969].

The values of $(\Delta G^\circ_{Na, Cl})'$ may also be compared, as in tables XVII–XXI. These range from 2.69 (Carcinus) to 2.73 kcal (Mytilus). Again, as in the case of potassium distribution, the results for the three species are in excellent agreement. For lobster (whole muscle), a much lower figure, 2.03 kcal, is found. For the intracellular phase (or phases), the comparable value is 3.19 kcal. These figures all indicate that skeletal muscle is not homogeneous, and that the values of ΔG°, $\Delta\mu^\circ$ and the other thermodynamic quantities may show considerable variation within the muscle. For many of the properties, however, values for whole muscle, averaged over the entire tissue, yield sufficiently accurate values to characterize the physicochemical and physiological properties of the whole tissue.

Standard Free Energies and Dissociation Constants

The standard free energy of potassium-sodium distribution is a measure of the deviation from the laws of perfect mixing, as represented by the classical Donnan equilibrium (chapter IV). In a two phase system in which the dielectric constant of each phase is 80, the change of standard chemical potential of each ion becomes zero. In that limiting case, the standard free energy of any ion exchange reaction, as referred to sodium, is zero. Then, according to equation 6.8:

$$(\Delta G^\circ_{K, Na})' = (\Delta\mu_K{}^\circ)' - (\Delta\mu_{Na}{}^\circ)' \tag{6.12}$$

where $(\Delta\mu_K{}^\circ)'$ and $(\Delta\mu_{Na}{}^\circ)'$ are defined by equations 6.4 and 6.5. In the limiting case of perfect mixing, $\Delta\mu_i{}^\circ$ is zero and α_i'' is 1.0. In that case, $(\Delta G^\circ_{i, Na})'$ is given by equation 6.10a. In solid or semi-solid phases, as referred to blood plasma, $(\Delta G^\circ_{K, Na})'$ is negative and (r_K) is greater than r_{Na}.

The deviation is caused by a difference in the dielectric constant of the structured water from that of the liquid phase water. This leads to high values of $\Delta\mu_k{}^\circ$ and $\Delta\mu_{Na}{}^\circ$. In the case of potassium, however, the intramuscular values of α_K'' are low. Therefore, the values of $(\Delta G^\circ_{K, Na})'$ for vertebrate and invertebrate muscle are usually of the order of -2.5 to -3.5 kcal. The relationship between $(\Delta G^\circ_{K, Na})'$ and ion distribution is as follows:

$$(\Delta G^\circ_{K, Na})' = -1.419 \log_{10} \frac{(r_K)}{(r_{Na})} \tag{6.13}$$

The constant 1.419 kcal is a value that applies to a temperature of 37 °C. As an approximation that applies roughly to temperatures from 25 to 37 °C, the value 1.4 kcal may be used. The following quantitative relationship between $(\Delta G^{\circ}_{K,Na})'$ and the ion distributions is then obtained:

$\frac{(r_K)}{r_{Na}}$	$(\Delta G^{\circ}_{K,Na})'$, kcal
1.0	0.0
5.0	– 0.98
10.0	– 1.40
50.0	– 2.38
100.0	– 2.80
200.0	– 3.22
500.0	– 3.78

For most of the values given in tables VIII–XXI, $(\Delta G^{\circ}_{K,Na})'$ falls within the limits of −2.6 to −3.2 kcal. In general, therefore, (r_K) exceeds r_{Na} by a factor of the order of 50 to 200. A summary of the standard free energies

Table XXII. Apparent standard free energies of electrolyte distribution

	K,Na	½ Ca	½ Mg,Na	Na,Cl
Mammals				
Man	–3.02	–0.84	–1.71	1.95
Pig	–3.05	–0.83	–1.77	2.10
Rabbit	–3.09	–1.03	–1.77	1.95
Rat	–3.05	–0.98	–1.75	2.03
Mean	–3.05	–0.92	–1.75	2.01
Cold-blooded vertebrates				
Frog	–2.92	–0.89	–1.26	1.73
Salmon	–3.12	–	–	2.01
Anguilla, tongue	–2.98	–1.03	–1.44	1.61
Anguilla, parietal	–3.19	–0.43	–0.89	1.92
Roman eel	–3.64	–1.20	–1.32	2.87
Mean	–3.17	–0.84	–1.23	2.03
Invertebrates				
C. maenas	–2.69	–1.05	–1.24	2.69
Pecten	–2.61	–	–	2.70
M. edulis	–2.62	–	–	2.73
Lobster, whole muscle	–2.82	–0.71	–1.26	2.03
Lobster, intracellular	–3.65	–1.38	–1.61	3.19
Mean	–2.87	–1.05	–1.37	2.67

of ion distribution is given in table XXII. For four species of mammals $(\Delta G^\circ_{K, Na})'$ varies only from -3.02 to -3.09 kcal, with a mean of -3.05 kcal. For cold-blooded vertebrates, the mean is -3.39 kcal, and the variability is much greater, ranging from -2.92 for the frog to -3.58 kcal for skeletal muscle of the Roman eel. For the invertebrates (Mytilus, Pecten, Carcinus), the values range from -2.61 to -2.69 kcal. The extreme value (-3.65 kcal) is found in the intracellular phase of lobster muscle. This differs considerably from the value of -2.82 kcal, as found for whole muscle. The values for whole muscle probably represent at least two or three different states of water. According to STENSEN's ideas [1662], these would be intrafibrillar, interfibrillar and membranous. The membranous phase, including connective tissue, would differ in electrolyte-water composition from the other two phases. Connective tissue, in general, shows values of $(\Delta G^\circ_{K, Na})'$ that are quite different from those of cellular phases [JOSEPH, 1971a, b].

When the values of $(\Delta G^\circ_{Na, Cl})'$ are compared for mammalian muscle, the mean value of 2.01 kcal is found. The values range from 1.95 (man and rabbit) to 2.10 kcal (pig). The variability is slightly greater than that found for $(\Delta G^\circ_{K, Na})'$. For cold-blooded vertebrates, the values of $(\Delta G^\circ_{Na, Cl})'$ are of the same order of magnitude as those for mammals, showing a mean of 2.03 as compared to 2.01 kcal. The individual values range from 1.61 (Anguilla tongue) to 2.87 kcal (Roman eel). For invertebrate muscle, the values of $(\Delta G^\circ_{Na, Cl})'$ are of a higher order of magnitude than those for mammals, showing a mean value of 2.67 kcal. Three species (Mytilus, Pecten and Carcinus) show remarkably similar values (2.73, 2.70 and 2.69 kcal). Lobster muscle deviates considerably, showing a value of 2.03 kcal for whole muscle and a value of 3.19 kcal for the intracellular phase.

The high values of $(\Delta G^\circ_{Na, Cl})'$ found in marine invertebrates reflect the very low values of both (r_{Na}) and (r_{Cl}) found in these tissues. This represents the effect of the intramuscular nitrogenous osmoregulatory substances, which lower the chemical potential of intracellular water to compensate for the high electrolyte concentration of the hemolymph. On the geological time scale, this kind of osmoregulatory functions has enabled such invertebrates to compensate for the long-range increase in salinity of the ocean. An extreme example of such an adaptation has been found in the case of *Ericheir sinensis,* which can live in both sea water and fresh water [GREEN, 1968]. Although its normal habitat is the rivers of China, it has adapted itself to the waters of Europe, and is said now to range from northern France to the Gulf of Finland. In fresh water, the blood has a freezing point of $-1.1\,^\circ C$, but in sea water the value is $-2.09\,^\circ C$, which is isotonic with that of the habitat. In fresh water the electrolyte metabolism thus seems to approach that of the vertebrate species, but in the ocean, it loses its osmoregulatory control of the hemolymph, resembling in this respect typical invertebrates such as Mytilus and Carcinus.

In intracellular osmoregulation, several amino acids play an important role [BRICTAUX-GREGOIRE *et al.,* 1962]. These are said to include alanine, glutamic acid, glycine, proline and threonine. Trimethylene oxide, an important end product of nitrogenous metabolism, also has been shown to have a significant osmoregulatory function.

With respect to the values of $(\Delta G^\circ_{\frac{1}{2}Ca,\,Na})'$ and $(\Delta G^\circ_{\frac{1}{2}Mg,\,Na})'$, there appear to be no remarkable differences among mammals, cold-blooded vertebrates and invertebrates. For calcium-sodium distribution, the values for mammals, cold-blooded vertebrates and invertebrates are, respectively, -0.92, -0.84 and -1.05 kcal. The most pronounced differences within any group are found in the lobster, where the intracellular values of both $(\Delta G^\circ_{Ca,\,Na})'$ and $(\Delta G^\circ_{Mg,\,Na})'$ are negative with respect to the values for whole muscle. As is the case of potassium and chloride distribution, the results reflect differences in the states of intrafibrillar, interfibrillar and membranous water.

Intramuscular States of Ions

The intracellular ionization constant, α_i'' for any ion is

$$\alpha_i'' = \frac{c_i''}{(c_i)''} \tag{6.14}$$

By means of equation 6.6, α_i'' is estimated, using calculated values of (r_i), $\Delta\mu_i$ and $\Delta\mu_i^\circ$. These values are given in tables VIII–XXI, and are summarized in table XXIII. In all tissues, the value of α_{Na}'' is taken as 1.0, assuming sodium to be completely ionized. Justification for this assumption will be given in later chapters. For the present, it may be stated that equations 6.1 and 6.2 are also based on that assumption. The values of $\Delta\mu_{Na}^\circ$ so calculated are in agreement with observed values of the resting electrical potential, E, and the action potential, E_a (equation 4.15). The theoretical explanation of potassium binding as opposed to the complete ionization of sodium will be given in chapter XI. An empirical explanation of the difference between the two ions was given by LOEB [1906].

Table XXIII. Ionization constants in muscle (values of α_i''/α_i')

	K	Ca	Mg	Cl
Mammals				
Man	0.023	2.3×10^{-5}	5×10^{-9}	0.020
Pig	0.030	10^{-5}	8×10^{-10}	0.017
Rabbit	0.020	1.6×10^{-5}	6×10^{-8}	0.028
Rat	0.021	3×10^{-5}	9×10^{-8}	0.039
Cold-blooded vertebrates				
Frog	0.004	6×10^{-5}	8×10^{-8}	0.011
Salmon	0.007	–	–	0.037
Anguilla, tongue	0.0036	5.4×10^{-5}	2×10^{-7}	0.059
Anguilla, parietal	0.010	10^{-3}	5.5×10^{-6}	0.10
Roman eel	0.009	7×10^{-6}	7×10^{-9}	0.158
Invertebrates				
C maenas	0.035	8×10^{-5}	2×10^{-8}	0.085
Pecten	0.038	–	–	0.073
M. edulis	0.036	–	–	0.102
Lobster, whole muscle	0.026	2.5×10^{-5}	3×10^{-8}	0.035
Lobster, intracellular	0.006	2.5×10^{-6}	10^{-9}	0.14

In table XXIII, ionic dissociation is expressed as α_i''/α_i'. This represents the ratio of the intracellular to the extracellular ionization constants. For potassium ion in mammalian muscle, the values range from 0.020 (rabbit) to 0.030 (pig), with an average of 0.024 for four species. In cold-blooded species, an average value of less than 0.01 is found. In anguilla tongue muscle and frog skeletal muscle the value is about 0.004, and in the Roman eel the value is 0.007, as compared with 0.01 for anguilla parietal muscle. In general, the group of cold-blooded vertebrates shows higher binding of potassium than that of the mammals (lower values of α_K''). The values are also low when compared with α_K'' of invertebrate muscle. These show an average value of 0.028 but the figures include the very low value of 0.006 for the intracellular phase of lobster muscle.

For the whole muscle of lobster, the value is 0.026, as compared with values of 0.035–0.038 for Mytilus, Pecten and Carcinus. These show values of -2.61 to -2.67 kcal for $(\Delta G°_{K,Na})'$, as compared with -2.82 kcal for the whole muscle of lobster and -1.65 kcal for the intracellular phase. High values of $\alpha_{K''}$ (weak binding) are thus correlated with low values (2.61–2.67 kcal) for $-(\Delta G°_{K,Na})'$. Strong binding ($\alpha_K'' = 0.006$) is connected with the value 3.65 kcal in intracellular lobster muscle. Accordingly, the varying values of α'' in invertebrates may well indicate the heterogeneous nature of muscle tissue, which contains intrafibrillar and interfibrillar water as referred to total water. In these species α_K'' is of the order of 0.004–0.01; these values are similar to that found in the intracellular phase of lobster muscle.

For intramuscular calcium, α_{Ca}'' is of the order of 10^{-5} or 10^{-6} in all vertebrate and invertebrate species. For mammals the average is 2×10^{-5} as compared with 4×10^{-5} for cold-blooded vertebrates. These differences are not great when one considers the great variety of species that are compared, ranging from scallop and crab to mammals, including man. One is compelled to recognize the great similarity of all species of muscular protoplasm, confirming the deep insight of HUXLEY's remarks as given in chapter III.

When the values of $\alpha_{Cl}''/\alpha_{Cl}'$ are compared, it is again found that the ionization constants are of similar orders of magnitude. The average values are 0.026 for mammals, 0.047 for cold-blooded vertebrates, and 0.087 for invertebrates. Thus invertebrates show the highest binding of chloride. If a reason is sought for this, it might be found in the high content of organic nitrogenous compounds that are commonly found in invertebrate muscle of these marine species. Such substances would be bound in part to the negatively charged macromolecular substances. This is especially true of those substances such as arginine and other basic amino acids or amines, which carry positive charges. Binding of organic cations, in which the tissues are rich, would induce a corresponding increase of osmotically inactive chloride ions.

It appears then, as a first approximation, that there are great similarities among all kinds of muscle tissue with respect to the free energies of ion distribution and intracellular binding of the ions. Since all the ions considered are of environmental origin, they enter only into reversible processes within the organism. Thus they conform to the principle of 'conservation of reversibility' that applies to phylogenetic and ontogenetic development. This principle tends to preserve the constancy of the *milieu intérieur* within any species and within each phylum. Since the ions that are involved are biologically invariant, long-term invariance or univariance of the physicochemical system is maintained by phase rule conditions of invariance and constraint. These conditions apply to the water and electrolytes of the *milieu intérieur,* which operate to maintain constant chemical potentials in organisms that are free to develop phylogenetically.

Chapter VII: The Heart

Electrolyte distribution in heart muscle is governed by the same laws that determine the chemical morphology of skeletal muscle and of other parenchymal and mesenchymal tissues. These are the thermodynamic conditions of invariance and constraint that, along with the genetic laws of phylogeny and ontogeny, control and channel the development of all cells and tissues of the body. Chemical potentials of electrolytes and water remain constant in blood plasma and other liquid phases of the *milieu intérieur.* These fluids undergo no changes of physicochemical state during the processes of growth, development and aging. Freezing points and electrolyte composition, as a first approximation, remain constant. On the other hand, the intracellular fluid of heart muscle, like that of skeletal muscle, exists in more than one state of aggregation. Therefore, electrolyte distribution does not conform to the classical Donnan equilibrium, which applies only to homogeneous solutions with the property of perfect mixing.

In the heterogeneous tissues of the mammalian heart, water exists in two or more states of aggregation. The myocardium undergoes periodic changes of state between diastole and systole. Therefore the system has at least one degree of freedom. This corresponds to a periodic and rhythmic univariant change of state of the muscle fibers, coordinated in a one-to-one relationship with contraction and relaxation of the fibers.

According to the principles that were developed in chapter VI, the normal electrolyte distribution of the mammalian heart may be described by the apparent standard free energies of potassium, calcium, magnesium and chloride distribution, as referred to sodium. These values depend on the values of $\Delta\mu_i^\circ$ for each of the five ions, and on the values of the ionization constants, α_i''. Values of the apparent standard free energies of distribution $(\Delta G^\circ_{i,Na})'$ and of α_i''/α_i' may be calculated from the water-electrolyte composition of heart tissue, according to the mathematical methods that were described in chapter VI. The results for human and pig heart tissue are presented in tables XXIV and XXV.

Table XXIV. Electrolyte balance in heart (man)[1]

	Na	K	Ca	Mg	Cl
Blood plasma, mEq	140	5	5	2.5	105
Tissue, mEq	69.9	80.4	–	16.0	55.1
Ratio (r_i)	0.50	16.1	–	6.4	0.52
RT ln (r_i), kcal	–0.43	1.72	–	1.15	–0.40
$\Delta\mu_i°$, kcal	0.54	0.54	–	1.08	–0.54
$\Delta\mu_i$	0.97	0.73	–	4.85	0.53
$(\Delta\mu_i°)'$	0.97	–1.18	–	–0.07	–0.14
$(\Delta G°_{i,Na})'$	–	–2.15	–	–1.00	0.83
α_i''/α_i'	1.10	0.045	–	0.00035	0.28

1 Widdowson and Dickerson [1960].

Table XXV. Electrolyte balance in heart (pig)[1]

	Na	K	Ca	Mg	Cl
Blood plasma, mEq	140	5	5	2.5	105
Tissue, mEq	55.5	106.1	4.0	23.3	38.8
Ratio (r_i)	0.40	21.2	0.80	9.3	0.37
RT ln (r_i), kcal	–0.57	1.80	–0.14	1.33	–0.61
$\Delta\mu_i°$, kcal	1.27	0.96	4.8	6.35	0.70
$\Delta\mu_i$	0.70	0.70	1.40	1.40	–0.70
$(\Delta\mu_i°)'$	1.27	–1.10	1.54	0.07	–0.09
$(\Delta G°_{i,Na})'$	–	–2.37	–0.50	–1.24	1.18
α_i''/α_i'	1.0	0.035	0.0044	0.00003	0.28

1 Widdowson and Dickerson [1960].

Thermodynamics of Electrolyte Distribution

The following formula is used to determine the values of $\Delta\mu_{Na}°$, the change of standard chemical potential of sodium ion between blood plasma and heart muscle:

$$\Delta\mu_{Na}° = \frac{RT\,[\Sigma c_i' - \Sigma (c_i)'']}{c_{Na}''} \qquad (7.1;\ 6.1)$$

This relationship was derived in chapter V as equation 5.25 and was used in chapter VI as equation 6.1 to estimate $\Delta\mu_{Na}°$ in skeletal muscle. In human blood plasma $\Sigma c_i'$ is taken as 315 mM/kg water. In heart muscle the sum of sodium, potassium and chloride concentrations may be estimated as 205.4 mM/kg water (table XXIV). Therefore:

$$\Delta\mu_{Na}^{\circ} = \frac{620 \times (0.315 - 0.205)}{0.0699} = 0.97 \text{ kcal} \tag{7.2}$$

The values of $\Delta\mu_i^{\circ}$ for the other four ions may be estimated from the corresponding values of z_i^2/b_i, as explained in earlier chapters. Then the respective values for potassium, calcium, magnesium and chloride ions are: 0.73, 3.65, 4.85 and 0.55 kcal.

In table XXV, the analytical results for the pig heart yield a value of 1.27 kcal for $\Delta\mu_{Na}^{\circ}$. This leads to the following values for the other ions (in kcal): potassium, 0.96, calcium, 1.80, magnesium, 6.35, and chloride, 0.70. Each value is higher than the corresponding value for human heart by a factor of the order of 30%. This implies that pig heart muscle has a lower dielectric constant than that of human heart.

The next step in elucidating the total electrolyte balance is the estimation of $\Delta\mu_{Na}$, according to the following equation:

$$\Delta\mu_{Na} = \Delta\mu_{Na}^{\circ} + RT \ln (r_{Na}) \tag{7.3; 6.2}$$

or, according to table XXIV for human heart:

$$\Delta\mu_{Na} = 0.97 - 0.43 = 0.54 \text{ kcal}$$

Therefore, the value of $\Delta\mu_K$ is also 0.54 kcal. The values of $\Delta\mu_{Ca}$ and $\Delta\mu_{Mg}$ (bivalent cations) are therefore 1.08 kcal, and $\Delta\mu_{Cl}$ is -0.54 kcal. This corresponds to a value of 0.54 kcal for δ, the equivalent change of chemical potential for each of the five ions. The respective values of $(\Delta\mu_i^{\circ})'$ may then be calculated from the relationships:

$$\Delta\mu_i = (\Delta\mu_i^{\circ})' + RT \ln (r_i) \tag{7.4}$$

The apparent change of chemical potential is defined as:

$$(\Delta\mu_i^{\circ})' = \Delta\mu_i^{\circ} + RT \ln \alpha_i''/\alpha_i' \tag{7.5}$$

The ratio of ionization constants α_i''/α_i' may then be estimated from the other parameters shown in tables XXIV and XXV. The values of the apparent standard free energies of distribution, $(\Delta G^{\circ}{}_{i,Na})'$ are also estimated from the data. The following equations are applicable:

$$(\Delta G^{\circ}{}_{K,Na})' = (\Delta\mu_K^{\circ})' - \Delta\mu_{Na}^{\circ} \tag{7.6}$$

Table XXVI. Electrolyte balance in heart (dog)[1]

	Na	K	Cl
Blood plasma (mM)	140	5	105
Tissue	47.5	104.0	39.6
Ratio (r_i)	0.34	20.8	0.38
RT ln (r_i), kcal	–0.67	1.87	–0.60
$\Delta\mu_i°$, kcal	1.85	1.39	1.01
$\Delta\mu_i$	1.18	1.18	–1.18
$(\Delta\mu_i°)'$	1.85	–0.69	–0.58
$(\Delta G°_{i,Na})'$	–	–2.54	1.27
α_i''	1.0	0.035	0.076

1 Darrow *et al.* [1939].

Table XXVII. Electrolyte balance in heart (cat)[1]

	Na	K	Cl
Blood plasma, mM	140	5	105
Tissue	60.0	96.4	46.6
Ratio (r_i)	0.43	19.3	0.44
RT ln (r_i), kcal	–0.52	1.83	–0.51
$\Delta\mu_i°$, kcal	1.16	0.87	0.58
$\Delta\mu_i$	0.64	0.64	–0.64
$(\Delta\mu_i°)'$	1.16	–1.19	–0.12
$(\Delta G°_{i,Na})'$	–	–2.35	1.04
α_i''	1.00	0.036	0.32

1 Darrow *et al.* [1939].

Table XXVIII. Electrolyte balance in heart (rat)[1]

	Na	K	Cl
Blood plasma, mM	140	5	105
Tissue	41.5	105.6	30.7
Ratio (r_i)	0.29	21.1	0.29
RT ln (r_i), kcal	–0.75	1.87	–0.75
$\Delta\mu_i°$, kcal	2.05	1.51	1.12
$\Delta\mu_i$	1.30	1.30	–1.30
$(\Delta\mu_i°)'$	2.05	–0.57	–0.55
$(\Delta G°_{i,Na})'$	–	–2.62	1.50
α_i''	1.0	0.035	0.068

1 Widdowson and Dickerson [1964].

$$(\Delta G^\circ_{1/2\mathrm{Ca,\,Na}})' = \tfrac{1}{2}(\Delta\mu_{\mathrm{Ca}}{}^\circ)' - \Delta\mu_{\mathrm{Na}}{}^\circ \tag{7.7}$$

$$(\Delta G^\circ_{1/2\mathrm{Mg,\,Na}})' = \tfrac{1}{2}(\Delta\mu_{\mathrm{Mg}}{}^\circ)' - \Delta\mu_{\mathrm{Na}}{}^\circ \tag{7.8}$$

$$(\Delta G^\circ_{\mathrm{Na,\,Cl}})' = (\Delta\mu_{\mathrm{Cl}}{}^\circ)' + \Delta\mu_{\mathrm{Na}}{}^\circ \tag{7.9}$$

The respective values are calculated from the values of $(\Delta\mu_{i^\circ})'$ as given in tables XXIV and XXV. The values of $(\Delta G^\circ_{\mathrm{K,\,Na}})'$ for human heart is -2.15 kcal, as compared with -2.35 kcal for pig heart. For $(\Delta G^\circ_{\mathrm{Na,\,Cl}})'$, the respective values are 0.83 and 1.18 kcal. For $(\Delta G^\circ_{½\mathrm{Mg,\,Na}})'$, the respective values are -1.00 and -1.24 kcal. The value of $(\Delta G^\circ_{½\mathrm{Ca,\,Na}})'$ for human heart is lacking. Thus, when comparison is possible, all the respective values of the apparent standard free energies of ion distribution are of similar orders of magnitude for a given pair of ions.

Thermodynamic constants for electrolyte balance in heart muscle for dog, cat and rat are presented in tables XXVI–XXVIII. Data for calcium and magnesium in these species are unavailable. From the available data for sodium, potassium and chloride ions, however, the following thermodynamic functions have been calculated: the apparent standard free energies of potassium-sodium and sodium-chloride distribution, and the ionization constants, α_i''/α_i' for potassium and chloride distributions. For $(\Delta G^\circ_{\mathrm{K,Na}})'$, the average value fo the three species is found to be about -2.50 kcal. This is not greatly different from the values -2.15 and -2.37 kcal, as calculated for man and pig in tables XXIV and XXV.

The values of α_K''/α_K' calculated in tables XXVI–XXVIII show an average of 0.36. This is not far from the values 0.45 and 0.35 found for human and pig heart (table XXIV and XXV). The average values of $\alpha_{Cl}''/\alpha_{Cl}'$ for dog, cat and rat is about 0.38. This may be compared with the values 0.25 and 0.28 for man and pig, calculated in tables XXIV and XXV.

Comparison of Heart and Skeletal Muscle

The values of the apparent standard free energies of ion distribution in five mammalian species are compared in table XXIX. Results are compared with those for skeletal muscle of four species of mammals. The results for heart muscle show an average value of -2.41 kcal for five species, ranging from -2.15 (man) to -2.62 kcal (rat). A value of -2.37 kcal is found in the pig. This is near the median value; the values of cat, dog and rat are, respectively, −2.35, −2.54 and −2.62 kcal. Thus the extreme range of variation is 0.47 kcal, which represents the difference in values for human and rat tissues.

In skeletal muscle the average value of $(\Delta G^\circ_{K,Na})'$ is −3.05 kcal. The limits of variation are much smaller than is found in heart muscle. It is represented by the values −3.02 and −3.09 kcal as found respectively in human and rabbit skeletal muscle. Corresponding to this narrow range of variation, there is a small standard error, of the magnitude of ±0.03 kcal. The standard error of $(\Delta G^\circ_{K,Na})'$ of heart muscle is 0.16 kcal. Thus the difference between the two tissues is highly significant.

Highly significant differences between heart and skeletal muscle are also found between the two groups of mammals when the values of $(\Delta G^\circ_{Na,Cl})'$ are compared. The mean value for heart muscle is 1.21 ± 0.25 kcal. The lowest value (0.83 kcal) is found for man, as compared with the upper limit of 1.52 kcal found in the dog. In four species of skeletal muscle, the mean value of $(\Delta G^\circ_{Na,Cl})'$ is 2.01 ± 0.07 kcal. As is the case with potassium-sodium distribution, the standard error in skeletal muscle is small when it is compared with heart. The difference of means (0.60 kcal) is much greater than the standard error in either group; this indicates a highly significant difference.

Small standard errors are also found in the values of $(\Delta G^\circ_{\frac{1}{2}Ca,Na})'$ and $(\Delta G^\circ_{\frac{1}{2}Mg,Na})'$ for skeletal muscle. For calcium-sodium distribution,

Table XXIX. Standard free energies of electrolyte distribution in mammalian muscle (values in kcal)

Species	K,Na	½Ca,Na	½ Mg,Na	Na,Cl
Heart muscle				
Man	−2.15	–	−1.00	0.83
Pig	−2.37	−0.50	−1.24	1.18
Dog	−2.54	–	–	1.52
Cat	−2.35	–	–	1.04
Rat	−2.62	–	–	1.50
Mean + SE	−2.41 ± 0.16	−0.50	−1.12	1.21 ± 0.18
Skeletal muscle				
Man	−3.02	−0.84	−1.71	1.95
Pig	−3.05	−0.83	−1.77	2.10
Rabbit	−3.09	−1.03	−1.77	1.95
Rat	−3.05	−0.98	−1.75	2.03
Mean + SE	−3.05 ± 0.03	−0.94 ± 0.10	−1.75 ± 0.03	2.01 ± 0.07

the mean value is -0.94 ± 0.10 kcal, and for magnesium-sodium distribution the mean is -1.75 ± 0.03 kcal. For magnesium distribution, the small differences among the four species may be quite insignificant. The differences among the four species with respect to calcium-sodium distribution are somewhat greater, but these may also be of small significance. Because only one or two species are represented for calcium and magnesium distribution in the series of values calculated for heart, no estimation of the standard error is possible. However, the value of $(\Delta G^\circ_{\frac{1}{2}\,Ca,\,Na})'$ for pig heart is -0.50 kcal. This is clearly beyond the standard error of the corresponding values for the group of four mammals for which the value in skeletal muscle is known. Thus the difference of 0.44 kcal between the two groups must be considered significant when the value -0.50 kcal is compared with -0.94 ± 0.10 kcal. Similarly, the difference between the values for $(\Delta G^\circ_{\frac{1}{2}Mg,\,Na})'$ in the two groups seems highly significant. This represents a difference between -1.12 kcal (the mean of human and pig heart) and -1.75 ± 0.03 kcal, which represents the mean of four species of mammalian skeletal muscle.

Therefore, it may be concluded that heart muscle differs significantly from skeletal muscle with respect to all four apparent standard free energies of distribution. As has been shown in preceding sections and in earlier theoretical studies [JOSEPH, 1971a, b], the standard free energies of ion distribution represent deviations from the laws of perfect mixing, as represented by the classical Donnan equilibrium. Thus for each of the five ions,

Table XXX. Ionization constants in muscle (values of α_i''/α_i')

Species	K	Ca	Mg	Cl
Heart muscle[1]				
Man	0.045	–	3.5×10^{-4}	0.28
Pig	0.035	0.0044	3.0×10^{-5}	0.28
Dog	0.035	–	–	0.076
Cat	0.036	–	–	0.32
Rat	0.035	–	–	0.068
Skeletal muscle[2]				
Man	0.023	0.000023	5×10^{-9}	0.020
Pig	0.030	0.00001	8×10^{-10}	0.017
Rabbit	0.020	0.000016	6×10^{-9}	0.028
Rat	0.021	0.00003	9×10^{-8}	0.039

1 Tables XXIV–XXXIII.
2 Tables VIII–XI.

the change of standard chemical potential, $\Delta\mu_i^\circ$, indicates a difference of dielectric constant between an intracellular phase and the *milieu inté rieur.* In skeletal muscle, $\Delta\mu_{Na}^\circ$ is of the order of 2.8 kcal. In human and pig heart, the calculated values of D″ are of the order of 45–50, as compared with values of the order of 30 calculated for skeletal muscle [JOSEPH, 1971a]. Accordingly, skeletal muscle shows higher deviations from the laws of perfect mixing than does heart muscle. This indicates a more highly ordered state of water in skeletal muscle.

In adult skeletal muscle, the water should be considered to be highly structured and in a state of very low entropy. In heart muscle, the water may be regarded as more loosely structured and in a state of higher entropy or relative disorder. In both tissues the intracellular dispersion medium differs greatly from the state of perfect mixing that is characteristic of true or perfect solutions, which are homogeneous and perfectly miscible. Values of $(\Delta\ G^\circ_{K,Na})'$ and the other standard free energies of distribution provide useful criteria as to the states of intracellular water. A value of zero indicates a homogeneous or unstructured system with a value of 80 for D″. Highly structured and well-ordered systems are characterized by values of the order of -3 kcal for $(\Delta G^\circ_{K,Na})'$ and by values of the order of 30 for D″.

Dependence of Ionization on State of Water

From the results given in tables XXIV–XXX, it is evident that the ionization constants of the various ions depend in some way on the intracellular state of water, as reflected by the standard chemical potentials of the ions and the standard free energies of distribution. For any ion, the change of standard chemical potential between an intracellular phase of dielectric constant D″ and a liquid phase of dielectric constant 80 is given by:

$$\Delta\mu_i^\circ = \frac{164\, z_i^2}{b_i}\left(\frac{1}{D''} - \frac{1}{80}\right) \tag{7.10}$$

This relationship has been given in earlier sections as equations 2.4 and 4.10. Then D″ can be calculated from $\Delta\mu_{Na}^\circ$

$$D'' = \frac{131}{1.64 + \Delta\mu_{Na}^\circ} \tag{7.11}$$

where 1.64 kcal is the calculated value of μ_{Na}° in water. The number 131 represents the product of 80, the dielectric constant of water by 1.64, the calculated value of $\mu_{Na}{}^{\circ}$ in kcal. The value of b_{Na} is taken as 1.25 A [LAIDLER and PEGIS, 1957]. In human heart, D″ is estimated as 50, as compared with the calculated value of 45 for pig heart (table XXXI). The comparable values in skeletal muscle are 29 (man) and 27 (pig). Values of $\Delta\mu_i{}^{\circ}$ for each of the other ions are estimated from the respective values of z_i^2/b_i, according to equation 7.10. In order to correlate the calculated values of $\Delta\mu_i{}^{\circ}$ with the ionization constants for the various ions, it is necessary, as a first approximation, to find a relationship between the standard chemical potential and the distribution of the ions between ionic and bound forms. Such an hypothetical relationship would be given in statistical mechanics as some form of a Boltzmann statistical distribution (chapter XI and XII). Provisionally, the following relationships may be tested for each ion, with the exception of sodium, for which α'' is taken as 1.0:

$$\Delta\mu_i{}^{\circ} + RT \ln \alpha_i''/\alpha_i' = 0 \qquad (7.12)$$

or

$$\alpha_i''/\alpha_i' = \exp\left(\frac{-\Delta\mu_i{}^{\circ}}{RT}\right) \qquad (7.13)$$

The values of α_i''/α_i' in table XXXI are compared with the values that have been estimated in tables XXIV and XXV for heart muscle of man and pig. In pig heart the values for calcium, magnesium and chloride ions are in fairly good agreement with those estimated in table XXV. By ‘fairly good agreement’ is meant at least the same order of magnitude. The values of 0.32 and 0.28 calculated for chloride may be considered to agree well; the same may be said for the values for magnesium (1.5×10^{-5} and 3×10^{-5}). For calcium, the values 4×10^{-4} and 44×10^{-4} may be considered to be of similar orders of magnitude. Thus equations 7.12 and 7.13 provide reasonable first approximations for the ionization constants. On the other hand, the values calculated for the ionization of potassium do not agree

Table XXXI. Ionization constants in human heart[1]

Ion	$\Delta\mu_i{}^{\circ}$, kcal	α_i''/α_i'	α_i''/α_i'
K	0.73	0.32	0.045
Ca	3.65	0.0025	–
Mg	4.85	0.0004	0.00035
Cl	0.53	0.61	0.28

Values of α_i''/α_i' in column 2 calculated from equations 7.12 and 7.13. Values in column 3 from table XXIV.

with those calculated in tables XXIV and XXV. The results differ by a factor of the order of 10. Thus the values calculated from $\Delta\mu_i°$ are 0.32 (man) and 0.22 (pig); the values from tables XXIV and XXV are 0.045 (man) and 0.035 (pig). Thus the ionization of potassium in heart muscle is only about 10% of that calculated from $\Delta\mu_K°$. It has been shown elsewhere that potassium is always highly bound, even in cells and tissues of relatively high dielectric constant. The reason may be that there are intracellular regions of low dielectric constant that are accessible to potassium but inaccessible to the other inorganic ions [JOSEPH, 1971a] (see also section in chapter VI, relating to the high binding of potassium in intramuscular phases of the lobster). The results may then indicate a heterogeneous structure in heart muscle, corresponding to a multi-state theory of water structure (chapter I).

The results for skeletal muscle of man and the pig are presented in table XXXII. In these tissues, the values of D″ are less than 30, and $\Delta\mu_{Na}°$ is 2.80 (man) and 3.12 kcal (pig). There are strong contrasts with the results of table XXXI. In skeletal muscle, fairly good agreement is found in the calculated values for potassium and magnesium, when α_i'' is estimated from

Table XXXII. Relation of dielectric properties to intracellular ionization in skeletal muscle

Ion	$\Delta\mu_i°$, kcal	(α_i''/α_i')	(α_i''/α_i')
Man[1] (D″ = 29)			
Na	2.80	1.0	1.0
K	2.10	0.033	0.023
Ca	10.50	4.1×10^{-8}	2.3×10^{-5}
Mg	14.0	1.4×10^{-10}	5×10^{-9}
Cl	1.54	0.082	0.020
Pig[2] (D″ = 27)			
Na	3.12	1.0	1.0
K	2.34	0.022	0.030
Ca	11.7	6×10^{-8}	10^{-5}
Mg	15.6	10^{-11}	8×10^{-10}
Cl	1.72	0.064	0.017

1 Values in column 2 calculated from equations 7.12 and 7.13. Values in column 3 from table VIII.

2 Values in column 2 calculated from equations 7.12 and 7.13. Values in column 3 from table IX.

$\Delta\mu_i^\circ$, and compared with the results of tables VIII and IX. Thus α_K''/α_K' as estimated from equations 7.10 and 7.11 are 0.023 (man) and 0.030 (pig).

For magnesium, the results are also of the orders of magnitude (10^{-9} or 10^{-10} in man; 10^{-10} or 10^{-11} in pig). The results for calcium in the two tissues, however, are of different orders of magnitude when the results of the two methods of calculation are compared. According to tables VIII and IX, values of $\alpha_{Ca}''/\alpha_{Ca}'$ of the order of 10^{-5} are obtained. These differ greatly from values of the order of 10^{-9} or 10^{-10}, estimated from $\Delta\mu_{Ca}^\circ$. Thus intracellular calcium in skeletal muscle is much more highly ionized than would be predicted from a simple Boltzmann distribution. The deviation may conceivably result from a competition with magnesium or other intracellular substances for binding sites. Chloride ion also deviates from a simple Boltzmann distribution. It differs from calcium in being more highly bound than would be predicted from equations 7.10 and 7.11. Thus the values estimated in tables VIII and IX are 0.020 (man) and 0.017 (pig). The comparable values estimated from $\Delta\mu_{Cl}^\circ$ are 0.082 (man) and 0.064 (pig). The orders of magnitude are not extremely different in the two cases, but there is no quantitative agreement, as found for potassium binding.

The results may be interpreted as indicating a highly heterogeneous and differentiated structure in skeletal muscle. In such a system, it may not be possible to assume a uniform homogeneous dielectric constant. Even when agreement is found between the two methods, as is the case with potassium binding in skeletal muscle, this may represent a fortuitous result in which α_K differs in various parts of the structure, and where the calculated values of α_K by either method represent only statistical means. In any case, the simple Boltzmann distribution, although a useful and suggestive hypothesis, would not represent the true picture in a system in which water may exist in several different states of aggregation.

Chapter VIII: Growth and Development

Growth and development of the mammalian or vertebrate organism are highly coordinated and well-ordered processes that are never completely independent of each other. The most rapid phases of both processes occur in the early period of life, especially during the period of rapid formation of bone. During the period of rapid growth and development, certain tissues undergo marked changes of chemical composition. This is especially true of skeletal muscle, when the water content decreases from a fetal level of about 90% to an adult level of about 75%. In this period, the sodium and chloride contents decrease to about one fourth the level of fetal muscle, or to about one fifth of the blood level. At the same time the potassium concentration increases by a factor of two, as referred to the fetal level. The highest concentrations are of the order of twenty to thirty times the blood level. The changes of electrolyte content are obviously not the independent variables. Each ion concentration is a well-ordered function of growth and development; the concentrations are related to the contents of water and protein, which bear a one to one relationship to the age of the organism. In muscle, for example, the chemical morphology and physicochemical state tend to be univariant functions of age and development [JOSEPH, 1971a].

In any tissue or cellular structure, chemical morphology and state develop embryologically and ontogenetically under well-defined conditions of invariance and constraint. Thermodynamic conditions of constraint are imposed by the invariant state of the *milieu intérieur.* In mammalian blood plasma, for example, the freezing point and chemical potential of water remain constant, as a first approximation, during the entire life span. This implies constant composition with respect to the inorganic electrolytes. According to phase rule principles, invariance of blood electrolytes and water imposes conditions of constant chemical potentials of electrolytes and water throughout the organism. These conditions stabilize the chemical composition of all cells and tissues by determining thermodynamic conditions of stability in the resting state of maximal accessibility. This is the state of normal homeostasis, which depends on a curve of normal growth and development. Intracellular chemical morphology is then dependent on growth and aging, under conditions of constraint that are imposed by the constancy of the *milieu intérieur.* Stability is imposed by reciprocal conditions of constraint. Thus it may be said that constant composition of somatic cells and tissues depends on the maintenance of a constant state of the *milieu intérieur.* However, it is also true that invariance of the body fluids depends on maintenance of constant states of the cells and tissues. Thus a significant displacement of the chemical morphology by nervous or hormonal influence would result in generalized systemic changes of the normal balance and changes of the properties of the body fluids. Such changes of chemical morphology with concomitant displacements of electrolyte balance may also occur in pathological states of sufficient severity.

Morphological development of cells and tissues in the normal animal is always a well-ordered process. This implies that the synthesis of the several intracellular or extracellular macromolecular substances is higly synchronized rather than independent. Normally, growth and development are processes that are univariant in time. Thus a tissue such as skeletal muscle develops with only one degree of freedom. This implies that at any point on the normal growth curve, there is only one standard state of maximal accessibility. During irreversible processes of growth or development, the standard state changes irreversibly with one degree of freedom. This implies that the set of n macromolecular polyelectrolytes develops in a highly constrained way that limits the entire set to one normal standard state at any given time. Therefore, the n components are synthesized as a well-ordered unit rather than independently. Independent synthesis of n components would imply a system of at least n degrees of freedom. This would constitute a mixture of high entropy and high disorder, in contrast to the normal well-ordered biological state of negative entropy. Maintenance of physiological states of maximal order implies a high coordination of normal growth and development with nutritional supplies of negative entropy and with an ordered state of energy and entropy balance at any given time.

Growth and Development of Skeletal Muscle

Water and electrolyte content in the skeletal muscle of post-partum rats at ages of 2–65 days was studied by HAZLEWOOD and NICHOLS [1969]. Concurrent measurements of the resting potentials were also observed. The electrical measurements may be compared with the earlier observations of FUDEL-OSIPOVA and MARTYNENKO [1969] on post-partum rats up to the age of 20 days. By the methods that were described in earlier sections, it is possible to calculate the resting potential from sodium, potassium and chloride content of the whole muscle, and to compare the calculated values with the measurements of HAZLEWOOD and NICHOLS and with those of FUDEL-OSIPOVA and MARTYNENKO. From the colligative properties of the tissue (chapter V), the values of the dielectric energy, $c_{Na}''\Delta\mu_{Na}^{\circ}$ and of the change of standard chemical potential of sodium may be estimated:

$$c_{Na}''\Delta\mu_{Na}^{\circ} = RT\,[\Sigma c_i' - \Sigma(c_i)''] \tag{8.1}$$

and

$$\Delta\mu_{Na}^{\circ} = \frac{RT\,[\Sigma c_i' - \Sigma(c_i)'']}{c_{Na}''} \tag{8.2}$$

For example, in the 16-day-old rat, the sum of internal sodium, potassium and chloride (Na + K + Cl) is 197.6 mM/kg water, and c_{Na}'' is 43.9 mM. Taking the value of RT as 620 cal at 37 °C,

$$c_{Na}''\Delta\mu_{Na}^{\circ} = 620\,(0.315 - 1.976) = 72.54 \text{ cal}$$
$$\Delta\mu_{Na}^{\circ} = 1.65 \text{ kcal}$$

and

$$\Delta\mu_{K}^{\circ} = 1.27 \text{ kcal}; \; \Delta\mu_{Cl}^{\circ} = 0.90 \text{ kcal} \tag{8.3}$$

These are obtained from the values of z_i^2/b_i for the various ions, taking the corrected radii as 1.25 Å for sodium, 1.66 Å for potassium, and 2.25 Å for chloride. The values of z_i^2/b_i are then approximately, 0.80 (Na), 0.60 (K) and 0.44 (Cl). The following relationships are used to determine $(\Delta\mu_{Na}^{\circ})'$, $(\Delta\mu_{K}^{\circ})'$, and $(\Delta\mu_{Cl}^{\circ})'$ from the data:

$$\Delta\mu_{Na} = \Delta\mu_{K}^{\circ} = -\,\Delta\mu_{Cl} \tag{8.4}$$

$$\Delta\mu_{Na} = \Delta\mu_{Na}^{\circ} + RT\ln(r_{Na}) \tag{8.5}$$

$$\Delta\mu_{K} = (\Delta\mu_{K}^{\circ})' + RT\ln(r_{K}) \tag{8.6}$$

$$\Delta\mu_{Cl} = (\Delta\mu_{Cl}^{\circ})' + RT\ln(r_{Cl}) \tag{8.7}$$

The values of $(\Delta\mu_{K}^{\circ})'$ and $(\Delta\mu_{Cl}^{\circ})'$ are calculated by combining the four equations and by using the observed values of the distribution ratios (r_{Na}), (r_k) and (r_{Cl}). The apparent standard free energies of distribution are then estimated from the relationships:

$$(\Delta G^\circ_{K,Na})' = (\Delta\mu_K{}^\circ)' - (\Delta\mu_{Na}{}^\circ)' \tag{8.8}$$

and

$$(\Delta G^\circ_{Na,Cl})' = (\Delta\mu_{Na}{}^\circ)' + (\Delta\mu_{Cl}{}^\circ)' \tag{8.9}$$

These values, calculated from the data of table XXXIII are given in table XXXIV. The values of the ionization constants α_K'' and α_{Cl}'' are estimated from the following equations:

$$(\Delta\mu_K{}^\circ)' = \Delta\mu_K{}^\circ + RT \ln \alpha_K''/\alpha_K' \tag{8.10}$$

and

$$(\Delta\mu_{Cl}{}^\circ)' = \Delta\mu_{Cl}{}^\circ + RT \ln \alpha_{Cl}''/\alpha_{Cl}' \tag{8.11}$$

The calculated values of α_K'' and α_{Cl}'' are presented in table XXXIV at various ages of rat skeletal muscle during the growth period. The resting potential, E, is also given in tables XXXIII and XXXIV. The values are calculated from the values of $\Delta\mu_{Na}$, by means of the equation:

$$FE + \Delta\mu_{Na} = 0 \tag{8.12}$$

Finally, the dielectric constant D″ is estimated from the relation:

$$D'' = \frac{131}{1.64 + \Delta\mu_{Na}{}^\circ} \tag{8.13}$$

where 1.64 kcal is the calculated value of $\mu_{Na}{}^\circ$ in water, and where 131 represents the product $164 \times 1/b_{Na}$, where $b_{Na} = 1.25$ Å.

According to the results of table XXXIV, the calculated value of $\Delta\mu_{Na}{}^\circ$ increases from 0.37 kcal in the 2-day-old rat to 2.72 kcal at the age of 32 days. The concurrent change of E (calculated) is from − 10 to − 76 mV. The corresponding observed change is from − 10.0 to − 74.1 mV. Agreement at the ages 4–16 days is less satisfactory (table XXXIII and XXXIV). However, there is also a lack of agreement between the values observed by HAZLEWOOD and NICHOLS and those found by FUDEL-OSIPOVA and MARTYNENKO. The question will be discussed in later sections.

From the data of tables XXXIII and XXXIV, it would appear that the adult values of E are observed between the ages of 16 and 32 days. According to FUDEL-OSIPOVA and MARTYNENKO, the adult level of −80 mV is observed at about 20 days. The calculated value of E is − 76 mV, corresponding to a value of 2.72 kcal for $\Delta\mu_{Na}{}^{\circ}$. This is in good agreement with the values given in tables VIII–XI for various species of adult mammals (table VI).

Table XXXIII. Growth and development of rat muscle[1]

	Age in days					
	2	4	8	16	32	65
Water, g	88.19	86.98	84.81	81.06	80.09	77.07
Na, mEq/kg water	112.4	105.0	70.3	43.9	30.1	24.1
K	72.5	76.5	95.4	118.6	128.5	145.3
Cl	61.8	58.9	50.0	35.1	24.6	16.3
Na + K + Cl	246.7	240.4	215.7	197.6	183.2	185.7
Na + K − Cl	123.1	122.6	115.7	127.4	132.0	153.0
$c_{Na}'' \Delta\mu_{Na}{}^{\circ}$, kcal	41.54	46.50	61.40	72.54	81.84	79.98
$\Delta\mu_{Na}{}^{\circ}$, kcal	0.37	0.44	0.87	1.65	2.72	3.32
$\Delta\mu_{Na}{}^{\circ}$, kcal	0.23	0.27	0.45	0.94	1.73	2.21
E, mV	−10.0	−11.7	−19.5	−40.8	−74.1	−95.9
Eq. wt., g	1,090	1,220	1,540	1,830	1,850	1,940
D''	68	63	52	40	30	26

1 Based on analytical results and observed electrical potentials [HAZLEWOOD and NICHOLS, 1969]. Comparison with calculated values of E is given in table XXXIV.

Table XXXIV. Development of rat skeletal muscle

	Age in days				
	2	4	8	16	32
$(\Delta G^{\circ})'$, kcal					
K,Na	−1.79	− 1.85	−2.22	−2.67	−3.09
Na,Cl	0.45	0.50	0.84	1.42	1.83
Ionization (α''/α')					
K	0.067	0.069	0.039	0.026	0.025
Cl	0.76	0.75	0.44	0.16	0.021
Electromotive force, mV					
E[1]	−10	−13	−36	−65	−76
E[2]	−10	−11.7	−19.5	−40.8	−74.1
E[3]	−24.9	−28.1	−52.5	−79.5	−80

1 Calculated from electrolyte composition, table XXXIII.
2 Observed, HAZLEWOOD and NICHOLS [1969].
3 Observed, FUDEL-OSIPOVA and MARTYNENKO [1964].

It is also evident that the adult levels of α_K'' and α_{Cl}'' are attained at about the same period. The potassium ionization constant reaches a final low value of 0.026 or 0.025 at the age of 16–32 days. The value of α_{Cl}'' arrives at the low value of 0.021 at 32 days. At 16 days the value is 0.18 so that, according to this criterion, the muscle fiber has not reached a full state of development. In adult man, the calculated values of α_K'' and α_{Cl}'' are in quantitative agreement with those of the 32-day-old rat. The values are respectively 0.023 and 0.020, as compared with 0.026 and 0.021 (table XXXIV and XXXV). For the pig, α_K'' is 0.030 and α_{Cl}'' is 0.017. Thus the ionization of both potassium and chloride are both age-dependent. For chloride, α'' in the pre-natal period may be of the order of 0.50 or higher, compared with the adult value of about 0.02. For potassium, the pre-natal value may be of the order of 0.07 or higher, as compared with an adult level of 0.020 to 0.030.

Table XXXV. Electrolyte balance in skeletal muscle

	Fetus	Newborn	Adult
Human skeletal muscle[1]			
K,Na	–1.28	–2.05	–3.02
Na,Cl	0.28	0.49	2.10
½ Ca,Na	0.00	–0.43	–0.84
½ Mg,Na	–0.62	–1.73	–1.71
Pig skeletal muscle[1]			
K,Na	–1.53	–2.75	–3.05
Na,Cl	0.28	1.53	2.10
½ Ca,Na	–0.23	–1.90	–0.83
½ Mg,Na	–0.53	–1.87	–1.77
Human skeletal muscle[2]			
K	0.070	0.067	0.023
Cl	0.50	0.045	0.020
Ca	0.20	0.0028	2.3×10^{-5}
Mg	0.04	3×10^{-6}	5×10^{-9}
Pig skeletal muscle[2]			
K	0.10	0.038	0.030
Cl	0.61	0.10	0.017
Ca	0.07	0.00025	1×10^{-5}
Mg	0.03	2.5×10^{-7}	8×10^{-10}

1 $(\Delta G^\circ)'$, kcal.
2 Ionization (α''/α').

Comparative values for calcium and magnesium can also be obtained for man and pig by the use of equations of the same form as 8.10 and 8.11. Here, $(\Delta\mu_{Ca}{}^{\circ})'$ and $(\Delta\mu_{Mg}{}^{\circ})'$ are estimated from $\Delta\mu_{Ca}$ and $\Delta\mu_{Mg}$, using observed values of (r_{Ca}) and (r_{Mg}) [JOSEPH, 1971a].

Both bivalent cations are fairly well ionized in the fetal period of both man and pig. The values of α'' range from 0.20 (Ca in man) to 8×10^{-10} (Mg in pig). In the fetus, D″ is 60 or higher, as compared to 30 or less in the adult man or pig. In the rat, D″ decreases from about 68 to 30 during the early post-partum period (table XXXIV). Thus α_{Ca}'' and α_{Mg}'' are much more sensitive to changes in D″ than are α_K'' and α_{Cl}''. This agrees with the fact that $\Delta\mu_{Ca}{}^{\circ}$ and $\Delta\mu_{Mg}{}^{\circ}$ are approximately five times the respective values of $\Delta\mu_K{}^{\circ}$ and $\Delta\mu_{Na}{}^{\circ}$. The values are in each case proportional to z_i^2/b_i for any ion. Therefore $\Delta\mu_i^{\circ}$ is very high for the bivalent ions, calcium and magnesium. According to the Boltzmann statistics, as explained in chapter VII, the formula

$$\Delta\mu_i^{\circ} + RT \ln \alpha_i''/\alpha_i' = 0 \qquad (8.14)$$

might serve as an approximation in certain limiting cases [JOSEPH, 1966b, 1971a, b].

Acid-Base Balance

Acid-base balance is dependent on the chemical morphology of cells and tissues. It is a function of the well-ordered synthesis of the *n* macromolecular substances. This determines chemical morphology and physicochemical state of cellular and extracellular structures as functions of growth, development and aging. Thus acid-base balance in the organism as a whole is always determined by the composition of each tissue with respect to the macromolecular polyelectrolytes. Electrolyte balance depends not only on the dielectric properties of each part of the system, but also on the fixed colloidal charge, which is determined by the nature and state of the polyelectrolytes.

In general, the intracellular and extracellular polyelectrolytes contain groups which combine with or dissociate hydrogen ions. Thus they may behave as acids, bases or amphoterie electrolytes which bear both postitive and negative charges. Intracellular polyelectrolytes include acids such as RNA and DNA, which are rich in ionized secondary phosphate. They also contain simple and conjugated proteins and esterified substances such as hexose phosphates. Basic substances include the protamines and histones, and certain kinds of nitrogenous derivatives. In general, the fixed charge of intracellular phases is negative; the value may be of the order of 100–150 mEq of net negative charge per kg water.

In extracellular phases such as the ground substance of connective tissue, the fixed colloidal charge is generally more variable than in intracellular phases. It may vary from about 30 mEq/kg water in tendon, to values that are higher than 200 mEq/kg water in articular or costal cartilage. Tendon is rich in the electrically neutral collagen, and has only a small content of the negatively charged glycoproteins and acid mucopolysaccharides, such as chondroitin sulfate. At the other extreme, cartilage contains many negatively charged carboxylate and sulfate groups. This leads to a highly negative charge density of the tissue as a whole.

The colloidal charge x depends on two factors: first, the mean equivalent weight of the cellular or extracellular colloids, and second, on the content of the cellular or extracellular structure with respect to colloid and water. The mean equivalent weight EW is defined as the number of grams of the mixture of colloids required to yield 1 Eq of fixed negative charge. This is measured as the net equivalents of inorganic cations necessary to neutralize the colloid, or the equivalents of hydrogen ions required to bring the colloid to its isoelectric point. Thus x is determined by the total mass of colloid and by the total number of titrable proton-binding groups between physiological pH and the isoelectric point. Therefore, x is determined by the nature and total quantity of the macromolecular polyelectrolytes. Water and electrolyte balance is determined by the same properties of biological structures. Electrolyte balance and water balance are never independent of acid-base balance. All balance is a function of growth, development and age.

The net negative fixed charge of extracellular and intracellular phases may be denoted as x, and expressed as Eq/kg of tissue water. The negative colloidal charge of any structure is neutralized by the inorganic cations, mainly sodium and potassium. In intracellular phases, most of the calcium and magnesium ions are in bound form. Therefore, as calcium or magnesium proteinates, they lower the net negative colloidal charge, as measured by x. As a first approximation:

$$\times + \Sigma(c_i)'' y_i = 0 \tag{8.15}$$

where y_i is 1 for potassium and sodium, and -1 for chloride ions. In general, since x is negative, there is a net excess of the univalent cations over the univalent anion, chloride. Thus if x is -100 mEq/kg water, the value of $\Sigma(c_i)''z_i$ is $+100$ mEq. In this approximation, calcium and magnesium are assumed to be highly bound, thus diminishing the net negative colloidal charge.

According to the above definitions:

$$x = (Na + K - Cl) \tag{8.16}$$

and

$$EW = \frac{W}{x} \tag{8.17}$$

where W is the total mass of the colloid in g/kg water. EW thus measures the mean equivalent weight of the colloids in g/Eq. It can be measured from the analytical figures for sodium, potassium and chloride content, expressed as Eq/kg water. As an approximation, W may be estimated from:

$$W = 1{,}000 - (H_2O)$$

where (H_2O) is the water content per 1,000 g of fresh tissue. Thus if a tissue contains 800 g of water per kg, W is taken as 250 g/kg water. In such a specimen (Na + K − Cl) is expressed as Eq/kg water. Thus when W is 250 g and x es 0.10 Eq, the value of EW is taken as 2,500 g/Eq. The mass of inorganic ions in such a specimen is generally of the order of 5 g. Referred to a value of 250 g for W, this represents an error of the order of about 2%, which will not be considered to be significant.

In table XXXIII, the mean equivalent weight EW is calculated for skeletal muscle in the growing rat between the ages of 2 and 65 days. At the age of two days, the calculated value is 1,090 g/Eq. The values increase rapidly for the first two weeks, reaching a level of 1,830 g at the age of 16 days. At the end of nine weeks (65 days), a value of 1,940 g is attained: this approaches the adult level that has been found for other skeletal muscle [Joseph, 1971a].

At the age of 2 days, the water content of rat skeletal muscle is 88.19 g/100 g fresh tissue. Thus the value of W is about 125 g/kg water. This value progressively increases to a level of 250 g/kg water at 32 days of age, and to about 300 g/kg water at 65 days. In adult skeletal muscle the water content is about 750 g/kg tissue. This corresponds to a value of W of about 330 g/kg water. Concurrently with the increases of W and EW, there is a progressive decrease of dielectric constant in rat muscle from 68 at 2 days to 30 at 32 days, and to 26 at 65 days. The change of standard chemical potential $\Delta\mu_{Na}^{\circ}$ increases from 0.27 to 2.72 and 3.32 kcal (32 and

65 days, respectively). The change of chemical potential, $\Delta\mu_{Na}$, increases from 0.23 to 1.73 and 2.21 kcal over the same time intervals. Proportional changes in the resting potential E occur during the same period. Calculated and observed values of E are shown in table XXXIV.

Concurrently with the changes of W, EW and E, there are changes in the calculated values of the ionization constants of potassium and chloride (table XXXIV). The value of α_K''/α_K' is 0.067 at 2 days and 0.060 at 4 days, decreasing to 0.039 at 8 days and to a final value of 0.026 or 0.025 at 16–32 days. Within the same period, $\alpha_{Cl}''/\alpha_{Cl}'$ decreases from 0.76 at 2 days to 0.021 at 32 days. At 16 days, however, the value is 0.16. Thus α_K'' reaches the adult level within 16 days, while α_{Cl}'' requires about 32 days to attain the final low value of 0.021. At 16 days the water content is 81.06% on a weight basis, as compared with 80.09% at 32 days. The respective values for EW are 1,830 and 1,850 g.

The most significant change is in the dielectric constant D″. This is about 40 at 16 days and about 30 at 32 days. The calculated value of $\Delta\mu_{Na}^{\circ}$ changes from 1.65 to 2.72 kcal during this period. Thus there are relatively large changes in the dielectric and electrical properties of the muscle between 16 and 32 days of age, during which period the changes of water content and equivalent weight are very small. It is evident that the physical properties may be more dependent on the state of aggregation of the intramuscular water than on its total quantity. This is also true of the ionization constants, α_K'' and α_{Cl}''.

It is thus evident that growth and development are not simply functions of colloidal weight, equivalent weight and water content. The thermodynamic and electrical properties are also functions of the state of aggregation of water and the colloidal macromolecules. This determines the dielectric constant of the dispersion medium, standard chemical potentials of the ions and the ionization constants. Thus states of aggregation may vary over certain time intervals during which the total weight of colloid W and the mean equivalent weight of the colloid remain nearly constant. Accordingly, D″ and x need not vary concurrently at every point on the growth curve. D″ depends on the state of aggregation of water, whereas x depends on the acid-base balance. In general, these are both functions of colloid-water content, which depends on the growth curve. However, over certain time intervals, x and EW may be constant, while D″, $\Delta\mu_i^{\circ}$, α_i'' and E approach the final adult values. The explanation may conceivably be that systems of constant polyelectrolyte-water composition may develop with respect to organization of the intracellular macromolecular aggregates.

The relationship between the total nitrogen content (g/kg water) and equivalent weight is shown in table XXXVI. A comparison is made between skeletal muscle and heart in man and pig. In human skeletal muscle total nitrogen increases from 12.3 (fetus) to 40.5 g/kg water (adult). The comparable increase in the pig is from 9.3 to 47.5 g/kg water. The equivalent weight increases from 1,150 to 2,140 g (man) and from 1,040 to 2,300 g in the pig. Dielectric constants decrease from 66 to 31 (man) and from 65 to 28 (pig). Thus in skeletal muscle there are parallel and concurrent changes in total protein, dielectric constant, water content and equivalent weights (or acid-base balance).

In heart muscle the change in all these parameters are comparatively small. The water content of heart decreases from 860 to 827 g in the human being and from 876 to 825 g in the pig. The lowest value of D″ in either

Table XXXVI. Acid-base balance and nitrogen content of skeletal muscle and heart

	Water, g	Total N	Eq. wt., g	D″
Skeletal muscle, man[1]				
Fetus				
13–14 weeks	907	12.3	1,150	66
22–22 weeks	807	16.7	1,210	62
Newborn	804	28.7	2,520	51
Adult	765	40.5	2,160	31
Heart muscle, man[2]				
Fetus	806	16.3	1,600	43
Newborn	841	23.3	2,170	50
Adult	827	27.7	2,200	50
Skeletal muscle, pig[1]				
Fetus, 46-day	916	9.3	1,040	65
Newborn	823	18.7	2,000	48
Four weeks	788	34.1	1,870	37
Adult	749	47.5	2,300	28
Heart muscle, pig[2]				
Fetus, 46-day	876	17.2	1,210	43
Newborn	830	25.1	1,670	47
Four weeks	820	29.0	1,670	47
Adult	825	31.7	1,750	44

1 Dickerson and Widdowson [1960].
2 Widdowson and Dickerson [1960].

species is 43 (fetus), as compared with 50 or 51 (newborn) and 44 or 50 in the adult. The changes of equivalent weight are also relatively small (1,600–2,200, human heart) and 1,210–1,750 (pig heart). It is evident that in heart muscle, the dielectric properties (states of aggregation(are relatively independent of age, while total nitrogen may increase by a factor of the order of 100%. This leads to increases of equivalent weight of the order of 30–50% and to rather small decreases of water content. In the case of heart muscle, there may be relatively large increases of nitrogen content and of equivalent weight, under fairly constant conditions with respect to the dielectric constant and standard chemical potentials. In heart tissue, large increases of protein nitrogen are not necessarily attended by correspondingly large changes in dielectric properties or electrolyte balance. Thus increases of total nitrogen of the order of 300 or 500% in heart muscle do not, as in skeletal muscle, lead to decreases of the dielectric constant or to marked changes in equivalent weight, acid-base balance or electrolyte balance. In this tissue, increases of protein content and decreases of water content may occur with no great changes of states of aggregation during the growth period.

Electrolyte balance in tissues such as skeletal muscle or heart depend essentially on two parameters, D″ and x. The change of standard chemical potential, $\Delta\mu_{i^\circ}$, of each of the five ions depends on D″. Ionization constants of the several ions are also related to D″. Thus the five values of $(\Delta\mu_i^\circ)'$ are functions of dielectric constant and the intracellular state of water. These are sufficient to determine the four values of the apparent changes of standard free energy $(\Delta G^\circ_{i,Na})'$ of each of the four ions, as referred to sodium. Thus there are four equations connecting the five distribution ratios: (r_{Na}), (r_K), (r_{Ca}), (r_{Mg}) and (r_{Cl}). A fifth relationship is required to establish the dependence of the intracellular concentrations: $(c_{Na})''$, $(c_K)''$, $(c_{Ca})''$, $(c_{Mg})''$ and $(c_{Cl})''$ and the constant extracellular concentrations: $(c_{Na})'$, $(c_K)'$, $(c_{Ca})'$, $(c_{Mg})'$ and $(c_{Cl})'$. The fifth relationship is given by the equation for acid-base balance.

Accordingly, there are five independent equations correlating the five intracellular ion concentrations, but these equations introduce two other variables or arbitrary parameters: D″ and x. If these can vary independently, the system would have two degrees of freedom. This, however, seems not to be the case. In normal growth and development, D″, x, W and EW are highly coordinated rather than independent functions of age. Therefore, the system as a whole develops with one degree of freedom. This is not to say that every parameter is an equally significant criterion of development. In certain periods, D″ may change significantly, while x, W and EW remain constant. In that case, states of aggregation may vary independently of colloid-water composition. The converse may also be true. In other periods or in other tissues (heart, for example), total nitrogen may change considerably at constant values of D″. In all such cases, however, the independent variable is the normal chemical morphology and physico-chemical state. Cells and tissues normally are univariant in time. They develop with only one degree of freedom.

Development of Heart, Liver, Kidney and Brain

Changes of water and electrolyte content of heart were determined in pre-natal, newborn, post-partum and adult periods of human beings and

Table XXXVII. Electrolyte balance in heart muscle[1]

	Fetus	Newborn	Adult
Human heart[2, 3]			
K,Na	–2.31	–1.95	–2.15
Na,Cl	1.17	0.89	0.83
½Ca,Na	–	–0.47	–
½Mg,Na	–	–0.87	–1.00
Pig heart[2, 3]			
K,Na	–2.47	–2.27	–2.37
Na,Cl	1.17	0.97	1.18
½Ca,Na	–0.93	–0.56	–0.54
½Mg,Na	–1.17	–0.83	–1.24
Human heart[2, 4]			
K	0.042	0.064	0.045
Cl	0.21	0.27	0.25
Ca	–	0.014	–
Mg	–	0.002	0.00035
Pig heart[2, 4]			
K	0.030	0.037	0.035
Cl	0.095	0.49	0.28
Ca	0.001	0.011	0.0044
Mg	3×10^{-6}	0.00008	3×10^{-5}

1 Widdowson and Dickerson [1960].

2 Human heart: 20- to 22-week fetus; pig heart: 46-day fetus.

3 $(\Delta G°)'$, kcal.

4 Ionization (α''/α').

pigs [WIDDOWSON and DICKERSON, 1960]. The results have been used to calculate the four apparent standard free energies of electrolyte distribution, $(\Delta G^\circ_{i,Na})'$ and the four ionization constants (K, Cl, Ca and Mg) (table XXXVIII). These eight parameters are compared in the fetal, newborn and adult states. Electrolyte balance in heart shows little tendency to change during pre-natal and post-natal growth periods. For example, the respective values of $(\Delta G^\circ_{K,Na})'$ in the fetal, newborn and adult states of human heart are −2.31, −1.95 and −2.12 kcal. The comparable values in pig heart are −2.47, −2.27 and −2.37 kcal. At any age, the difference between the two species is of the order of 0.2 or 0.3 kcal. There is no marked aging tendency in either species.

This also appears to be true of $(\Delta G^\circ_{Na,Cl})'$ in the fetal heart of both species the value is 1.17 kcal. The values of $(\Delta G^\circ_{\frac{1}{2}Ca,Na})'$ and $(\Delta G^\circ_{\frac{1}{2}Mg,Na})'$

Table XXXVIII. Electrolyte balance in liver[1]

	Fetus	Newborn	Adult
Human liver[2]			
K,Na	−2.37	−2.18	−2.50
Na,Cl	0.69	0.76	0.95
½Ca,Na	−0.27	−0.45	−0.46
½Mg,Na	−1.55	−1.03	−1.20
Pig liver[2]			
K,Na	−2.30	−2.37	−2.55
Na,Cl	0.20	0.96	1.23
½Ca,Na	−0.44	−0.31	−0.56
½Mg,Na	−1.18	−1.14	−1.35
Human liver[3]			
K	0.027	0.044	0.026
Cl	0.79	0.30	0.38
Ca	0.089	0.066	0.013
Mg	0.0023	0.00033	0.00016
Pig liver[3]			
K	0.031	0.030	0.028
Cl	0.28	0.18	0.25
Ca	0.040	0.030	0.0027
Mg	0.001	0.00036	1.8×10^{-5}

1 WIDDOWSON and DICKERSON [1960].
2 $(\Delta G^\circ)'$, kcal.
3 Ionization (α''/α').

are also of similar magnitudes in the two species, and show no marked aging changes. Skeletal muscle, on the other hand, shows strong progressive changes from the fetal to the adult state when the values of $(\Delta G°)'$ are compared. In human skeletal muscle, for example, $(\Delta G°_{K,Na})'$ shows values of -1.28, -2.05 and -3.02 kcal, as it develops from the fetal to the adult state. The comparable values for the pig are -1.53, -2.75 and -3.05 kcal (table XXXV). Similar results are found in the development of rat muscle from 2 to 32 days (table XXXIV). Here $(\Delta G°_{K,Na})'$ changes from -1.79 to -3.09 kcal, and $(\Delta G°_{Na,Cl})'$ changes from 0.28 to 2.10 kcal. The same values are found in pig muscle, which differs only in the newborn state. Large changes of $(\Delta G°_{Ca,Na})'$ and $(\Delta G°_{\frac{1}{2}Mg,Na})'$ are also observed in skeletal muscle. In heart, however, the changes are comparatively small and non-progressive.

This is also true of the various values of the ionization constants, α_i'', in human and pig heart. For potassium, α_K'' changes only from 0.042 to 0.045 between fetal and adult states, as compared with values of 0.030 and 0.035 for pig heart. In human heart, α_{Cl}'' changes only from 0.21 to 0.25, as compared with values of 0.098–0.28 in the pig. Magnesium is highly bound in all states of the pig heart; the values of α_{Mg}'' range from 3×10^{-5} to 3×10^{-6}. Calcium is more highly ionized, showing values of α'' ranging from 0.001 to 0.011. Both calcium and magnesium are more highly ionized in human heart, but there is no evidence of any strong aging tendency. In skeletal muscle, on the contrary, both α_{Ca}'' and α_{Mg}'' change greatly in pre-natal and post-natal growth periods. In the human fetal state, for example, α_{Ca}'' is 0.20 and α_{Mg}'' is 0.04. The respective adult levels are 2.3×10^{-5} and 5×10^{-9}. Thus α_{Ca}'' decreases by a factor of about 10^4 and α_{Mg}'' decreases by a factor of about 10^7. No comparable tendency is found in heart muscle.

In human and pig liver, the four values of $(\Delta G°)'$ in each species are, with one exception, of similar magnitudes. Also, with one exception, the values show little or no tendency to change from the fetal to adult states. The exceptional case is that of $(\Delta G°_{Na,Cl})'$, which changes from a level of 0.20 (fetus) to 1.23 kcal (adult). The other values of $(\Delta G°)'$ are quite similar to those found in heart muscle of both species. Little aging tendency is found in either tissue.

The values of α_i'' in liver are also of similar magnitudes to those found in heart, and for the most part, show only small or moderate changes with age. The values of α_K'' show almost no change; in the human fetus the value is 0.027, as compared with 0.026 in the adult. In the pig, the comparable values are 0.031 and 0.028. Also in the pig, α_{Cl}'' changes only from 0.28 to 0.25, as compared with a change from 0.79 to 0.38 in human liver. Rather large decreases are found for α_{Mg}'' both in human and in pig liver. In human liver the value decreases from 0.0023 to 0.00016, a factor of the order of 15. In pig liver, the change is from 0.001 to 1.8×10^{-5}, a factor of more than 50. These changes are of much smaller magnitudes than those found in skeletal muscle, where α_{Mg}'' decreases by a factor of the order of 10^7.

Kidney and brain, according to the results of tables XXXIX and XL, more nearly resemble heart muscle than liver, with respect to growth and development. The values of $(\Delta G°)'$ are similar for each kind of ion in each tissue, and no value shows a well-defined aging tendency. For example, the values of $(\Delta G°_{K,Na})'$ given in kcal are as follows in the adult tissues:

Table XXXIX. Changes of electrolyte balance during growth and development of kidney[1]

	Fetus	Newborn	Adult
Human kidney[2]			
K,Na	−2.00	−1.87	−1.83
Na,Cl	0.57	0.54	0.37
½ Ca,Na	−0.43	−0.38	−0.37
½ Mg,Na	−0.74	−0.80	−0.64
Pig kidney[2]			
K,Na	−1.61	−2.38	−2.00
Na,Cl	0.39	0.52	0.60
½ Ca,Na	−0.28	−0.26	−0.34
½ Mg,Na	−0.59	−0.83	−0.80
Human kidney[3]			
K	0.050	0.065	0.048
Cl	0.38	0.36	0.62
Ca	0.028	0.032	0.062
Mg	0.0024	0.0019	0.0089
Pig kidney[3]			
K	0.090	0.060	0.048
Cl	0.54	0.80	0.62
Ca	0.010	0.013	0.062
Mg	0.0071	0.0088	0.0089

1 WIDDOWSON and DICKERSON [1960].
2 $(\Delta G°)'$.
3 Ionization (α''/α').

– 1.83 (human kidney), – 2.00 (pig kidney), – 2.32 (human brain, – 2.20 (pig brain). The respective values of $(\Delta G^\circ_{Na, Cl})'$ taken in the same order for adult kidney and brain are: 0.37, 0.60, 0.85 and 0.76 kcal. Similarly, the values of $(\Delta G^\circ_{\frac{1}{2}Ca, Na})'$ and $(\Delta G^\circ_{\frac{1}{2}Mg, Na})'$ are respectively of similar magnitudes when adult kidney or brain are compared with respect to calcium or magnesium distribution. Kidney and brain are quite comparable with respect to any value of $(\Delta G^\circ)'$ when the tissues are in the adult state. They differ greatly from both skeletal muscle and liver both with respect to the magnitudes of $(\Delta G^\circ)'$ and with respect to aging tendencies of electrolyte distribution. In most properties they more nearly resemble heart than skeletal muscle or liver.

Table XL. Electrolyte balance in brain[1]

	Fetus	Newborn	Adult
Human brain[2]			
K,Na	–2.02	–1.82	–2.32
Na,Cl	0.59	0.39	0.85
$\frac{1}{2}$Ca,Na	–0.42	–0.24	–0.42
$\frac{1}{2}$Mg,Na	–0.82	–0.63	–0.95
Pig brain[2]			
K,Na	–1.78	–2.46	–2.20
Na,Cl	0.46	0.59	0.76
$\frac{1}{2}$Ca,Na	–	–	–
$\frac{1}{2}$Mg,Na	–0.75	–1.10	–0.92
Human brain[3]			
K	0.043	0.068	0.031
Cl	0.73	0.39	0.44
Ca	0.060	0.030	0.016
Mg	0.0062	0.0070	0.00014
Pig brain[3]			
K	0.070	0.040	0.036
Cl	0.49	0.32	0.73
Ca	–	–	–
Mg	0.0041	0.00068	0.018

1 WIDDOWSON and DICKERSON [1960].
2 $(\Delta G^\circ)'$, kcal.
3 Ionization (α''/α').

Relationships between Electrolyte Balance and Body Weight or Size

Accurate growth curves for rats over extensive periods of time are available from the results of McCay *et al.* [1935, 1939]. Recent measurements of electrolyte balance, electrical potentials and body weight during the first two months of life have been made by Hazlewood and Nichols [1969]. The results show a high correlation between the observed potentials in rat muscle and the growth curve. It was found that the resting potentials were related to age by the formula:

$$V(t) = 80 - 102\, e^{-0.105(t)} \tag{8.18}$$

where t is the age in days, V(t) is the resting potential expressed in millivolts as a function of t, and e is the base of natural logarithms. The value of V(t) approaches a limit of about −80 mV at about 35–45 days, when the body weight is from 100 to 150 g. Expressing the resting potential V as a function of body weight, Hazlewood and Nichols found:

$$V(BW) = 80 - 62\, e^{-0.042(BW)} \tag{8.19}$$

where (BW) is expressed in grams. This equations shows that the limiting or asymptotic value of −80 mV is reached at body weights greater than 100 – 150 g. The relationship between the resting potential and body weight was found to be highly significant. According to the results of tables XXXIII and XXXIV, limiting values of the dielectric constant D″ are attained at about the same time (32 days). At that age, the calculated and observed values of E are of the order of −75 to −80 mV, and D″ is about 30. Thus the dielectric constant or internal states of aggregation of polyelectrolyte complexes appear to depend on age and weight in much the same way as do the electrical properties.

The relationship between age and several other properties that are related to electrolyte balance are shown for a number of other cells and tissues in table XLI. These properties refer to dielectric constant, equivalent weights and dielectric energy, expressed as cal/kg water, and calculated as $c_{Na}{}''\Delta\mu_{Na}{}^{\circ}$. The tissues that are studied include skeletal muscle, heart, liver, brain and kidney of man and the pig.

From the fetal to the adult state of human or pig skeletal muscle, the values of D″ decrease from about 65 to about 30. This is about the same magnitude as that found in rat skeletal muscle (table XXXIII). Somewhat similar changes are found in human and pig liver, but the aging changes

Table XLI. Growth and development of human and pig tissues

Tissue	Skeletal muscle	Heart	Liver	Brain	Kidney
Dielectric constant					
Man					
Fetus	66	43	64	58	54
Newborn	51	50	56	58	57
Adult	31	50	49	58	65
Pig					
Fetus	65	43	54	57	57
Newborn	48	51	52	56	63
Adult	28	44	43	57	58
Equivalent weight, g/Eq					
Man					
Fetus	1,140	1,600	2,070	1,370	1,480
Newborn	2,520	2,170	3,430	1,880	1,670
Adult	2,160	2,200	3,640	2,370	2,490
Pig					
Fetus	1,040	1,210	2,160	1,210	1,010
Newborn	2,000	1,660	2,250	1,450	1,580
Adult	2,300	1,750	2,730	2,460	2,720
Dielectric energy, cal/kg water					
Man					
Fetus	36.58	73.78	32.24	49.49	61.96
Newborn	71.30	74.40	53.32	61.24	60.05
Adult	80.60	68.20	59.52	53.82	40.19
Pig					
Fetus	43.40	70.68	55.86	60.05	54.55
Newborn	71.92	63.86	56.42	61.72	42.10
Adult	83.70	71.30	67.58	52.63	60.05

are of smaller orders of magnitude than those found in skeletal muscle. No comparable aging changes of D″ are found in heart, brain or kidney.

All the tissues show increases in the mean equivalent weight during the periods of growth and development. The changes in skeletal muscle, brain and kidney appear to be of about the same orders of magnitude in man and pig. The fetal values of EW in the two species appear to be of the order of 1,000 – 1,400 g/Eq in muscle, brain and kidney. The adult values are usually of the order of 2,000–2,500 g/Eq in the three tissues. Thus the equivalent weights may be said to increase by factors of the order of 2 during the pre-natal and post-partum growth periods.

In liver, on the other hand, the fetal values of EW are of the order of 2,000 g, and the human adult levels are higher than 3,500 g. These values are much higher than those found in skeletal muscle, brain or kidney. Heart differs from the other tissues in showing rather small increases in EW. In man the increase is from 1,600 (fetal) to 2,200 g (adult). In pig the corresponding values are 1,210 (fetus) and 1,750 g (adult). The increases are of the order of 50 rather than 100%, as found in skeletal muscle.

With respect to dielectric energy, the most significant increases are found in skeletal muscle. In man the value of $c_{Na}''\Delta\mu_{Na}^{\circ}$ increases from the fetal level of 36.58 kcal/kg water to the adult level of 80.60. In the pig the respective values for fetus and adult are 43.40 and 83.70 cal. No comparable systematic changes are found for any of the other four tissues. In heart, the values generally range from about 65 to 75 cal/kg water. In the other tissues the values are generally less than 65 cal, and in kidney they may be as low as 40 or 45 cal. In human fetal liver, the value is 32.24 cal, and D″ is 64. In this tissue, intracellular water appears to approach the liquid state. However, in the newborn and adult states, the respective values of dielectric energy are 53.32 and 59.52 cal/kg water. These values are of similar magnitudes to those in pig liver and other adult tissues.

In general, it may be concluded that the colloidal composition of cells and tissues usually changes appreciably during the periods of pre-partum and post-partum growth and development. This is shown by the increases of equivalent weight that are usually found in these periods. These changes reflect the changing composition of the cellular polyelectrolytes, including proteins and nucleic acids. In certain tissues, especially skeletal muscle, these changes are attended by striking concurrent changes in the dielectric energy, and other electrical properties, all of which may be regarded as functions of body weight. In other tissues such as heart, brain and kidney, the dielectric properties tend to remain constant. Therefore, they are relatively independent of age and weight.

If chemical morphology is regarded as the primary property on which all physicochemical and physiological functions depend, then it may be said that cells and tissues differ greatly with respect to the variations of chemical morphology with age and body weight. Skeletal muscle changes greatly in all the secondary properties that are dependent on chemical morphology. Brain and kidney are examples of tissues in which the primary and secondary properties are less dependent on age.

Electrolyte Metabolism and Growth

According to views expressed by LOEB [1906], electrolyte balance in all types of cells, animal or plant, is a function of macromolecular synthesis or growth. '... the potassium is used in the building up of more complex compounds in which the K cannot be dissociated as a free ion. If a tissue utilizes one kind of metal in this way, for example, K, while another metal, for example, Na, is chiefly used for the formation of dissociable compounds with Na as a free ion, the consequence will be that the ashes of a tissue contain K and Na in altogether different proportions from what they are contained in the surrounding solution. I think we may take it for granted that, at least, K forms a non-dissociable constituent of the protoplasm of a number of tissues of animals and plants.'

The existence of extensive classes of organic macromolecules that bind potassium at the expense of sodium has recently been reported [CHRISTENSEN *et al.*, 1971]. These synthetic macrocyclic compounds typically contain 'central hydrophilic cavities ringed with either electronegative or electropositive binding atoms and exterior flexible frameworks exhibiting hydrophobic behavior. They show a pronounced ability to bind a wide variety of cations or anions and in many cases, to undergo marked conformational changes during binding.'

Selective ion-binding properties are shown by a wide variety of macrocyclic compounds, including polyethers, polyamines, polythioethers and many other types, with ring structures containing 9–60 total atoms, 3–20 oxygen atoms and 1–4 attached hydrocarbon rings. The possibility of these substances serving as models for naturally occurring biological analogues has been suggested by CHRISTENSEN *et al.* [1971]. Intracellular analogues having the properties of anion or cation binding of these macrocyclic substances could possibly be synthesized during the growth periods of animals or plants. These properties would account rationally for electrolyte metabolism as a univariant function of growth and macromolecular synthesis, as postulated by LOEB.

According to the results of table XXXIV, α_K''/α_K' in rat skeletal muscle decreases from a level of 0.067 at 2 days to 0.025 at 32 days. Over the same period, $(c)''$ increases from 72.5 to 128.5 mM/kg water, as the water content decreases from 88.19 to 80.09 g/kg muscle. The corresponding values of $(c_{Na})''$ decrease from 112.4 to 30.1 mM. Skeletal muscle thus behaves as a system in which the electrolyte-water composition has the properties of a univariant set. This shows that electrolyte balance is a univariant function of growth (or protein synthesis). Since, according to LOEB, potassium forms a non-dissociable constituent of the protoplasm of plants, as well as of animals, all electrolyte metabolism is a function of growth [JOSEPH, 1971b]. Plants synthesize intracellular and extracellular substances by absorption of photochemical energy. Animals obtain energy from proteins, carbohydrates and fats in the diet. Normal electrolyte balance and metabolism in animals require nutrient supplies of the essential minerals. Plants are dependent on the proper biotic and edaphic conditions of the soil. Animal cells grown in artificial culture media require mineral electrolytes in the correct proportions [EAGLE, 1965].

Plants belong to a form of life defined as *vie oscillante* [BERNARD, 1878]. 'In the variable state *(vie oscillante)*, the living organism is not entirely subordinate to external conditions, yet it is so bound up in them that it is subject to all their variations... We cannot therefore admit in living organisms a free vital principle which fights against external conditions.' BERNARD's principle, which is one of homeostasis agrees with CARNOT's principle (second law of thermodynamics) that an invariant system can do no work [JOSEPH, 1071b]. In the mammalian organism *(vie constante)*, the cells of the body possess no internal mechanisms to resist fluctuations in the internal environment. No internal 'vital forces' or mechanisms are present to control intracellular state in the face of extracellular changes.

Chapter IX: Free Energy and Work

As has been shown in chapter VIII and in earlier sections, many cells and tissues of the body are characterized by high levels of configurational free energy. This is especially true of mammalian skeletal muscle and heart. In the adult state, the dielectric energy of such tissues is of the order of 80 cal or 350 J/kg water when measured as $c_{Na}''\Delta\mu_{Na}^{\circ}$ or as RT $[\Sigma c_i' - \Sigma(c_i)'']$. Thus electrolyte distribution is determined by the change of standard chemical potential of sodium ion, $\Delta\mu_{Na}^{\circ}$, which in turn depends on the dielectric constant, D'', and on the intracellular state of water, as it is organized by the colloidal aggregates. In adult skeletal muscle, D'' is of the order of 30, as compared with the values of 40 or 50 that are found in mammalian heart muscle at various ages (chapters VI–VIII).

In the contraction of skeletal muscle, the configurational free energy may be quantitatively converted to work [JOSEPH, 1971a]. In the human arm muscles, the water content is of the order of 250 g [HILL, 1944]. Therefore, the available dielectric energy is of the order of 20 cal, or 85 J. This agrees well with the measurable values of about 90 J estimated by HILL for the work of the arm muscles in lifting a ball of 300 g to a height of 30 m. This yields a figure of 9 kg/m for the work, or about 88 J. This value, converted to kinetic energy, yields a fairly accurate estimate for the speed with which a baseball (150 g) can be delivered by a professional player [JOSEPH, 1971a]. Thus the maximal work of the biceps and brachialis muscles may be estimated from the configurational free energy, as determined by the term RT $[\Sigma c_i' - \Sigma(c_i)'']$. This is also proportional to the action potential, E_a, and to the change of standard chemical potential, $\Delta\mu_{Na}^{\circ}$. In human skeletal muscle, $\Delta\mu_{Na}^{\circ}$ is about 2.8 kcal, corresponding to an action potential of about 120 mV. The sodium concentration is about 0.028 mol/kg water. Therefore the dielectric energy ($c_{Na}''\,\Delta\mu_{Na}^{\circ}$) is calculated as about 78.4 cal or 330 J/kg water. Thus electrolyte distribution, action potential and muscular work are highly correlated and well-ordered functions of dielectric energy or configurational free energy [JOSEPH, 1971a].

In other cells and tissues besides skeletal muscle and heart, there are considerable quantities of dielectric energy, usually of the order of 200–300 J/kg water. In skeletal muscle, about 350 J/kg water are quantitatively converted to external work in a maximal response. In heart muscle, the conversion factor is much smaller (of the order of 2% or less). In other tissues such as liver, kidney and brain, the dielectric energy appears to represent other forms of irritability or responsiveness than mechanical contraction or tension. According to classical conceptions of cellular irritability [GLISSON, 1654; VON HALLER, 1757], cells and tissues respond in specific ways to a variety of stimuli. As shown in chapter I, this is explicable on a bi- or multi-state theory of cellular or tissue water, but is inexplicable on a uni-state theory.

In heterogeneous systems, the coexistence of two or more phases in different states of aggregation is explained by differences in configurational free energy or entropy, leading to differences in standard chemical potentials and standard free energies of ion distribution.

No such differences are possible in homogeneous systems with perfect mixing of water and electrolytes. In such systems, the macromolecular polyelectrolytes are in solution, and the differences of configurational free energy and entropy tend to vanish. Therefore, the physicochemical state approaches that of the classical Donnan equilibrium, in which water occurs as a homogeneous liquid with a dielectric constant that is of the order of 80. In such a system, $\Delta\mu_i^\circ$ of each kind of ion approaches zero. This is also true of $(\Delta G^\circ_{K,Na})'$ and the other standard free energies of ion distribution. Such systems are characterized by null values of configurational free energy, configurational entropy and dielectric energy. In a system with only one possible state of water, changes of state of water, colloids and electrolytes would not be possible. Therefore, the properties of responsiveness and irritability would disappear, and there would be no possibility of external mechanical work produced by internal configurational changes of state.

Action Potentials and Muscular Work

According to equations developed in chapter IV, the action potential, E_a or $-E^\circ$, the resting potential, E and spike potential E_{Na} are related in the following way:

$$-FE = -FE^\circ + RT \ln r_{Na} \tag{9.1}$$

where F is the Faraday constant (96,500 C or 23,060 cal/V/Eq).

Also:

$$FE^\circ + \Delta\mu_{Na}^\circ = 0 \tag{9.2}$$

and

$$FE_{Na} + RT \ln r_{Na} = 0 \tag{9.3}$$

The action potential E_a is equal and opposite in sign to E°. Therefore:

$$- FE = FE_a + RT \ln r_{Na} \tag{9.4}$$

In muscular physiology, $- RT \ln r_{Na}$ is proportional to the 'spike potential', E_{Na} (equation 9.3). This is usually of the order of 30–40 mV in mammalian or frog muscle. The action potential, E_a, is usually of the order of 120 mV, as compared with typical values of −80 mV for E. Values of E, E_{Na} and E_a in mV can be converted to the respective values of

$$\Delta\mu_{Na}, -\frac{RT}{F} \ln r_{Na}$$

and $\Delta\mu_{Na}^\circ$, as expressed in kcal. In this conversion, −100 mV corresponds to 2.306 kcal, and 1 kcal corresponds to 43.4 mV. Therefore, the typical values of resting potential, action potential and spike potential may be converted to thermal units (calories) or work units (joules) in the following way:

100 mV = − 2.306 kcal = − 9,650 J.

Resting potential:

− 80 mV = 1.84 kcal = 7.72 J.

Action potential:

− 120 mV = 2.76 kcal = 11.58 J.

Spike potential:

40 mV = − 0.92 kcal = − 3.86 J.

As has been shown in chapters VI and VIII, calculated values of E and E° (or E_{Na}) are also of the above orders of magnitude. It follows that the calculated values of E_{Na} or $(E + E_a)$ are also of the correct orders of magnitude.

According to the results of earlier sections, the dielectric energy or configurational free energy of the system is related to the electrolyte balance by the formula:

$$c_{Na}''\Delta\mu_{Na}^{\circ} = RT\,[\Sigma c_i' - \Sigma(c_i)''] \tag{9.5}$$

In adult human beings, $\Sigma c_i'$ may be taken as approximately 0.315 mol/kg water, and the intramuscular value of $\Sigma(c_i)''$ may be taken as 0.185 mol/kg water. At 37 °C, the value of RT is about 620 cal. Therefore:

$$c_{Na}''\Delta\mu_{Na}^{\circ} = 620 \times 0.13 = 80.6 \text{ cal/kg water}$$

$$80.6 \times 4.183 = 337 \text{ J/kg water.}$$

In general, the dielectric energy of adult mammalian muscle is of the order of 350 J/kg water, or 83.7 cal/kg water [JOSEPH, 1971a]. The dielectric energy of skeletal muscle can also be estimated from the formula:

$$c_{Na}''\,\Delta\mu_{Na}^{\circ} = -\,96{,}500\;c_{Na}''\,E^{\circ}\ (J) \tag{9.6}$$

Taking the value of c_{Na}'' as 0.028 mol/kg water, and E_a as 120 mV,

$$\begin{aligned}\text{Dielectric energy} &= 96{,}500 \times 0.028 \times 0.120\\ &= 325 \text{ J/kg water.}\end{aligned}$$

This is in good agreement with the value 337 J/kg water, calculated as above from the difference $[\Sigma c_i' - \Sigma(c_i)'']$. Therefore, the dielectric energy corresponds to the configurational free energy of electrolyte distribution. It is also related to the dielectric constant, D″, which is in turn related to the intracellular states of aggregation of the *n* macromolecular components and of water.

In growth and development, skeletal muscle behaves as a system of (n + 6) intracellular substances with one degree of freedom in respect to age. Since it can exist in either of two limiting morphological states, resting and stimulated (or relaxed and contracted), it has an additional degree of physiological activity. The magnitude of the external work of skeletal muscle per kg water is proportional to the decrease of configurational free energy that occurs in contraction. It is also, according to equation 9.6, proportional to c_{Na}'' and to the action potential, E_a. The action potential is therefore determined by a change of configurational free energy of the fiber or system of fibers rather than by changes of membrane permeability or ion transport. It is thus related to a change of order, disorder, and mixing properties of water, electrolytes and contractile proteins within the fiber.

Muscle irritability, tension and work can be explained only on the basis of a bi- or multi-state theory of intracellular water. It cannot be explained on the basis of a uni-state theory of water, on which the classical Donnan equilibrium is based.

A uni-state theory implies a state of homogeneity and perfect mixing of water, electrolytes and colloidal polyelectrolytes within the cell. Thus it does not consider internal changes of chemical morphology, energy and entropy. On the contrary, actual systems contain protoplasm which can exist in more than one state of aggregation, and which is always structured, well-ordered and heterogeneous. Protoplasm cannot be described therm-dynamically as a system with perfect mixing. Its description requires a theory of configurational free energies of distribution based on phase rule conditions of invariance and constraint. Such a theory is provided by GIBBS' 'equilibrium of heterogeneous substances' rather than by the theory of 'membrane equilibrium' postulated by DONNAN on the basis of a uni-state theory of water homogeneity [JOSEPH, 1971a].

Dielectric Properties of Muscle

According to equations 9.1, 9.2 and 9.4, the action potential, E_a (or $-E°$) is directly related to the change of standard chemical potential of sodium, $\Delta\mu_{Na}°$. It is therefore directly related to the dielectric constant D″. The relationship is:

$$D'' = \frac{131}{1.64 + \Delta\mu_{Na}°} \tag{9.7}$$

where 131 represents the product 80×1.64 where 80 is the dielectric constant of pure water, and where 1.64 kcal is the calculated value of $\mu_{Na}°$ in water. The value is calculated from the BORN equation [1920], as modified by LAIDLER and PEGIS [1957]. In table XLII, observed values of E_a for adult

Table XLII. Electrical potentials and dielectric constant of skeletal muscle

Species	Observed potentials E	E_a	E_{Na}	$\Delta\mu_{Na}°$ kcal	D″	References
Frog	–88	124	36	2.74	30	NASTUK and HODGKIN [1950]
	–92.8	130	37	2.99	28	NASTUK [1951]
	–91.8	125	33	2.85	29	ADRIAN [1956]
	–85	112	27	2.58	30	NICHOLLS [1956]
Cat	–79.5	116	36	2.67	30	TRAUTWEIN *et al.* [1953]
Guinea pig	–84.5	121	36	2.79	29	TRAUTWEIN *et al.* [1953]

cat, guinea pig and frog muscle are compared with values of the resting potential E and the spike potential, E_{Na}. Calculated values of $\Delta\mu_{Na}{}^{\circ}$ and D″ are also given. Four independent determinations of E_a yield values ranging from 112 to 130 mV. The corresponding values of D″ range from 23 to 30.

Similar values are found for E_a in mammalian muscle (cat, 116 mV; guinea pig, 121 mV). The values of $\Delta\mu_{Na}{}^{\circ}$ are 2.67 kcal (cat) and 2.79 kcal (guinea pig), yielding respective values of D″ of 30 and 29. These may be compared with calculated values for the 32-day-old rat, as given in chapter VIII, table XXXIII, ($\Delta\mu_{Na}{}^{\circ}$, 2.72 kcal, D″, 30). The value of E_a calculated from these figures is:

$$E_a = 2.72 \times 43.4 = 118 \text{ mV}$$

This agrees well with the figures for frog, cat and guinea pig.

Table XLIII shows values of $c_{Na}{}''$ $\Delta\mu_{Na}{}^{\circ}$, $\Delta\mu_{Na}{}^{\circ}$, E_a and D″ calculated from analytical results for skeletal muscle given in chapter VI, tables VIII–XXI. The following species are included – man, pig, rabbit and rat (table VIII–XI); cold-blooded vertebrates – frog, salmon, Roman eel, *Anguilla anguilla* (table XII–XVI); invertebrates – *Carcinus maenas,* Pecten, *Mytilus edulis* and *Nephrops norvegicus* (table XVII–XXI).

For four species of mammals, the results for $\Delta\mu_{Na}{}^{\circ}$ are 2.81 (man), 3.12

Table XLIII. Calculated dielectric properties of skeletal muscle

	Dielectric energy $c_{Na}{}''^{2}/_{Na}{}^{\circ}$, cal/kg water	$\Delta\mu_{Na}{}^{\circ}$, kcal	D″	E_a,mV[1] mV
Mammalian muscle				
Man	80.60	2.81	30	122
Pig	83.70	3.12	28	135
Rabbit	74.23	2.68	30	116
Rat	77.40	2.58	31	112
Cold-blooded vertebrates				
Frog	68.85	2.35	33	102
Salmon	72.22	2.57	31	112
Anguilla tongue	99.57	2.16	34	94
Anguilla parietal	82.20	2.05	35	89
Roman eel	65.00	2.60	31	113
Invertebrate				
Carcinus maenas	144.7	2.68	30	116

(pig), 2.68 (rabbit), and 2.58 kcal (rat). The respective values of D″ are 30, 28, 30 and 31, agreeing well with the figures of table XLII.

For cold-blooded vertebrates, the analytical figures yield the following values in kcal for $\Delta\mu_{Na}{}^{\circ}$: 2.35 (frog), 2.52 (salmon), 2.50 (parietal muscle of anguilla), 2.16 (tongue muscle of anguilla) and 2.60 (Roman eel). For *C. maenas* the value is 2.68 kcal. The value for *C. maenas* is obtained from the resting potential. For the other invertebrates, values of $\Delta\mu_{Na}{}^{\circ}$ cannot be observed or calculated; the values are assumed or hypothetical. The corresponding values of D″ for cold-blooded vertebrates and invertebrates, as given in table XLIII are 33 (frog), 31 (salmon), 35 (parietal of anguilla), 34 (tongue of anguilla), 31 (Roman eel). These values tend to be somewhat higher than those for mammalian muscle, but are clearly of the same orders of magnitude. The value of D″ for the muscle of *C. maenas* is 30, agreeing well with the values calculated for mammalian muscle.

Values of the dielectric energy, $c_{Na}{}'' \Delta\mu_{Na}{}^{\circ}$, may also be calculated from the data given in Chapter VI. These values as given in table XLIII are of the following orders of magnitude: 80–85 cal (mammals), 60–100 cal (cold-blooded vertebrates), 145 cal *(C. maenas)*. The high value in Carcinus muscle is a result of the high intracellular concentration of sodium (54 mM/kg water). High intramuscular sodium is also characteristic of the other marine invertebrates described in chapter VI, tables XVII–XXI. Thus in species in which D″ is of the order of 30, and in which $\Delta\mu_{Na}{}^{\circ}$ is of the order of 2.8 kcal, intracellular sodium concentrations of 50–80 mM would yield values of the order of 150–225 cal for the dielectric energy. This type of muscle, due to the high sodium content of sea water, would be characterized by very strong tensile properties as measured in g/cm^2.

It has been shown in an earlier study that the maximal work of muscular contraction can be approximately estimated from the dielectric energy of human skeletal muscle [JOSEPH, 1971a]. HILL [1944] has shown that in lifting a ball of 300 g weight to a height of 30 m, a man uses about 9 kg/m or 88 J of energy. Taking the weight of water in the biceps and brachialis muscles of an adult man to be approximately 250 g, this indicates that in a maximal effort, the arm muscles can deliver 36 kg/m or 352 J/kg muscle water to an external object such as a cricket ball [HILL, 1952] or a baseball [JOSEPH, 1971a]. As has been shown in equation 9.6, this is in fair agreement with the dielectric energy, as calculated for an action potential of 120 mV, when the muscle sodium is 28 mEq/kg water. The agreement would be nearly exact for an action potential of 120 mV when $c_{Na}{}''$ is 30 mEq/kg water.

It has been shown in chapter I (equation 1.15) that in the athermal phase of muscular contraction:

$$\Delta G_c = -T\Delta S_c \tag{9.8}$$

where ΔG_c is the change of configurational free energy and ΔS_c is the increase of configurational entropy. Accordingly, the energy transformations are expressed by:

$$\Delta G_c = -W_{max} \tag{9.9a}$$

and

$$T\Delta S_c = W_{max} \tag{9.9b}$$

where W_{max} is the maximal work of contraction. If this is estimated as 350 J or 83.7 cal/kg water, the entropy change ΔS, at 37 °C is 0.27 entropy units.

$$\Delta S = \frac{83.7}{310} = 0.27 \text{ cal/deg./kg water}$$

In the human arm muscles, which contain only 250 g water, the entropy change is 0.0675 entropy units. Thus the external work results from an increase of configurational free energy due to a change of state of intramuscular water. In passing from the extended state to the contracted state of the myofibrils, there is in increase of dielectric constant from the unstimulated level D″ to the stimulated level of about 80. This results in a calculated *decrease* of standard chemical potential of sodium of 2.76 kcal/Eq. For an action potential of 120 mV (1 kcal = 43.4 mV):

$$\Delta\mu_{Na}{}^{\circ} = \frac{120}{43.4} = 2.77 \text{ kcal}$$

$$= (\mu^{\circ}{}_{Na})_{resting} - (\mu^{\circ}{}_{Na})_{contracted}$$

Hence

$$W_{max} = 0.03 \times 2{,}770 = 83.1 \text{ cal/kg water}$$

$$= 348 \text{ J/kg water.}$$

In these calculations, very rapid contraction is assumed (within a few msec after stimulation). The general relationship between ΔG_c, ΔH_c and $T\Delta S_c$ is:

$$\Delta G_c = \Delta H_c - T\Delta S_c \tag{9.10}$$

In a very rapid isothermal, reversible contraction, when the entropy change, ΔS_c, is quantitatively converted to work with nearly 100% efficiency, ΔH_c may approach zero (equation 5.8). In the recovery from such a process, the myofibrils return very rapidly to the initial state. This requires a supply of negative entropy from intracellular nutritional sources. According to the preceding calculations, this amounts to – 0.27 entropy units/kg or to 0.0675 units/250 g muscle water. The value 83.1 J, as given above for the maximal work very nearly corresponds to 20 kcal for a water content of 250 g in the human arm muscles. It also corresponds to a decrease of 0.0675 entropy units.

The reversible extension of the muscle fibers therefore requires a nutritive source of negative entropy. In the anaerobic phase of muscular contraction, there is a breakdown of 1 molecule of glucoe to form 2 molecules of lactic acid. Considering only the initial and final states of this process:

Glucose = 2 lactic acid

$$\Delta G = \Delta H - T\Delta S$$
$$\Delta G = -29.88\ \text{kcal};\ \Delta H = -17.67\ \text{kcal};\ T\Delta S = 12.21\ \text{kcal}$$

In the maximal work of the arm muscles, $T\Delta S_c = 20$ cal. To restore the system to its initial state of low entropy requires the quantitative conversion of 20/1,221 mol of glucose to lactic acid. This amounts to 0.00155 mol of glucose. In the process, the ratio of $T\Delta S$ to ΔH is about 70%. The ratio of $T\Delta S$ to ΔG is about 40%.

In the repolarization of the muscle fibers to the initial state, the low configurational entropy of the entire system is restored. This signifies that water, electrolytes, and contractile proteins are restored to the initial configuration of the stretched fibers. The system is again in a state of high irritability, *tonus,* and configurational free energy. It is again in a position to deliver rapid external work that is quantitatively related to changes of dielectric energy or configurational free energy, and of high efficiency when referred to the cyclical process of contraction, recovery, and the compensating process of anaerobic glycolysis.

In these various processes, it is seen that the actual muscular contraction depends on ΔG and $T\Delta S$, which are quantitatively related (equation 9.8). In this process (corresponding to electrical depolarization), external work is related to changes of configuration of the contractile protein fibers rather than to exothermic or endothermic chemical reactions, involving the breaking of chemical bonds (as in the exchanges of phosphate bond energy).

Changes of configuration of the protein fibers involves breaking and rearrangements of the protein-water aggregates, but these processes may involve only $T\Delta S$ rather than ΔH. The same is true of the changes of state of sodium and the other ions, which change with respect to standard chemical potentials. Here also, the free energy changes appear to involve only entropy terms rather than changes of enthalpy.

Chemical bond energy appears to be involved mainly in the repolarization or recovery process rather than in the excitation or depolarization processes that accompany stimulation and contraction. In the glycolytic reactions and their accompanying side reactions, bonds are broken. The independent variable should be regarded as the change of state of water that accompanies the breakdown of the well-ordered state of low entropy that is characteristic of the standard state of stretched fibers. This may be regarded as the one degree of freedom in the system that determines all secondary properties. A change of state of water necessarily implies changes of state of electrolytes and all other solutes including nutrients, metabolites and gases. In a system with one degree of freedom, the initial and final states would be well-ordered: therefore all changes of secondary properties, since they depend only on the change of chemical morphology would also be well-ordered. Therefore, they are primarily determined by an increase of entropy of the entire system, rather than by exothermic or endothermic changes of enthalpy. It would be logical to regard all *chemical* changes involving breaking or synthesis of chemical bonds to be functions of morphology and state, rather than the contrary. In all active physiological processes, morphology and state eventually return to the standard state, characterized by negative configurational entropy. This is a function of growth, development and aging. Maintenance of normal irritability and *tonus* thus requires a continuous supply of negative entropy of nutritional origin. This implies irreversible exothermic reactions such as glycolysis, formation of carbon dioxide and water, with simultaneous transfer of heat to the environment and transfer of negative entropy to the well-ordered intracellular and extracellular macromolecular structures.

It is evident then that the common determinant of all secondary physiological and physicochemical properties is the intracellular state of water, as determined by chemical morphology, nutrition and respiration. The secondary properties include resting and action potentials, membrane permeabilities, electrical conductance and impedance and all other transport phenomena. These would change simultaneously with the state of water in a well-ordered system with one degree of freedom.

Athletic Performance

According to what has been said in preceding, the arm muscles of an athlete are capable of delivering about 88 J of energy to an external object [HILL, 1944]. The weight of a baseball, as used in the professional game, is about $5\frac{1}{4}$ oz. or 150 g. If 88 J are imparted to this object, as it is thrown horizontally, the velocity, v, is given by the formula:

$$v = \frac{2 \times 88 \times 10^7}{150}$$

where 10^7 is the number of ergs in 1 J. The calculated velocity is 34.2 m/sec. or about 2 km/min. This is approximately 1.3 mile/min, a figure which corresponds to the best performances of pitchers in professional baseball.

It would seem to be possible to estimate other maximal athletic performances on the basis of the dielectric energy or configurational free energy of human skeletal muscle. For example, the water content of the flexor muscles of the human thigh is of the order of 2.5 kg, as compared with 250 g for biceps and brachialis of the arm. Converting to kg/m as the energy unit, 900 J amounts to 90 kg/m as the available dielectric energy. Let us assume that in the high jump, the best athletes are able to transfer this energy into the vertical displacement of the body. Then a man weighing 76.7 kg would be able to lift his center of gravity 90/76.7 or 1.17 m above its vertical standing position. Assuming the resting center of gravity to be 1 m above ground level, the highest position would then be 2.17 m, or 84 in (7 ft). This is a very creditable performance for a first rate high jumper, although it is not quite at the level of world records.[1]

In the 100 m sprint, the athlete attains a maximal velocity of about 10 m/sec. Comparing the maximal kinetic energy with the total dielectric energy of the flexor muscles of the thigh (90 kg/m), one obtains:

1 Values of mass of water in the flexor muscles of the thigh can be estimated from the measurements of the physical characteristics of high jumpers in the Olympic games [TANNER, 1964]. The average height of high jumpers was found to be 191.5 cm, and the average weight was 76.7 kg. Subischial leg length was found to be 86.0 cm, and the circumference of the thigh was found to be 57.2 cm. Assuming the flexor muscles to account for half the total volume, and the water content to be 75% by weight, the intramuscular fluid of the flexor muscles would be of the order of 2.6 kg. Taking the value of 35 kg/m/kg water as the dielectric energy, the value is 2.6×34 or 91 kg/m. This is approximately ten times the value estimated for the biceps and brachialis muscles of the arm [HILL, 1964].

$$v^2 = \frac{s \times 90 \times 10^8}{\frac{1}{2}m}$$

where m is the mass of the body (70,000 g), and where s is the number of strides that are necessary to attain the maximal velocity, v. If this is taken as 10 m/sec (or 1,000 cm/sec), then:

$$s = \frac{10^6 \times 35{,}000}{10^7 \times 900} = 4$$

Thus at least 4 strides are required to attain a velocity of 10 m/sec, and this represents a maximal kinetic energy of 3,500 J. The maximal velocity is attained only after perhaps one tenth of the 100 m distance has been completed. The high jump also requires a preliminary sprint. It is evident that maximal kinetic energy can never be attained from a standing position.

Isometric Tension

Skeletal muscle develops mechanical tension along lines of force, the directions of which are determined by the anatomical arrangement of the myofibrils. The energy of an elastic fiber is proportional to its tension, t, and length, l [MEYER, 1951–1952]. Thus:

$$dE = T\,dS + t\,dl \tag{9.11}$$

and

$$dG = -\,S\,dT + l\,dt \tag{9.12}$$

where E is energy, G is free energy, and S is entropy. When the chemical potentials μ_i'', are taken as independent variables:

$$-\,S\,dT - l\,dt = m_{H_2O}''d\mu_{H_2O}'' + \Sigma m_i''d\mu_i'' \tag{9.13}$$

At constant temperature:

$$-\,l\,dt = m_{H_2O}''d\mu_{H_2O}'' + \Sigma m_i''d\mu_i'' \tag{9.14}$$

or

$$-\,.018\,l\,dt = d\mu_{H_2O}'' + .018\,\Sigma c_i''d\mu_i'' \tag{9.15}$$

In this set of equations, P and V are replaced by t and l, where these become the conjugate variables for mechanical work.

A growing muscle has two degrees of freedom, the first of which represents chemical composition as a function of time. The second is represented by the contractile properties. Both degrees of freedom are related to respective morphological states, contracted and extended.

In a simple physiological response, it may be assumed that the composition of any fiber or system of fibers remains constant and that the energy of contraction depends on the dielectric energy or configurational free energy, which changes upon stimulation. Dielectric energy is given by the product $c_{Na}''\Delta\mu_{Na}^{\circ}$. In differential form:

$$l\,dt = \Sigma\alpha_i''(c_i)''\,d\mu_i'' \qquad (9.16)$$

Except during a very short period in which there are transient changes of pressure and volume due to an increased chemical potential of water, it is assumed that μ_{H_2O} is constant. It is assumed that the values of α'' for four kinds of ions (K, Ca, Mg and Cl) are small, and that α_{Na}'' is 1.0. Long-term effects due to fatigue are not considered. If l and c_{Na}'' remain constant:

$$l\,dt = c_{Na}''d\mu_{Na}^{\circ} \qquad (9.17)$$

where $d\mu_{Na}^{\circ}$ is taken as positive. An increase of tension occurs when the dielectric energy decreases due to an increase of dielectric constant. The high energy of sodium ions is then transferred to the contractile proteins after a transient increase of pressure in the fibers. For unit length and cross sectional area:

$$t = 10.20\,\bar{d}\,w \times c_{Na}'' \qquad (9.18)$$

where t is expressed in g/cm. The constant 10.20 is a conversion factor between g/cm of mechanical energy and joules. Tension is expressed as t, and $\Delta\mu_{Na}^{\circ}$ is given in joules. The density of the fibers is expressed as $\bar{d}$ and the water content per g of fiber is denoted as w. When w is taken as 800 g water per kg muscle, and when $\bar{d}$ is taken as 1.05:

$$t = 8.6\,c_{Na}''\Delta\mu_{Na}^{\circ} \qquad (9.19)$$

When muscle contracts, the dielectric constant increases, and t is positive. In adult mammalian muscle, c_{Na}'' is about 0.03 mol/kg water, and the change of standard chemical potential of sodium is about 2.8 kcal, or 11,712 J. Then:

$$t = 8.6 \times 0.03 \times 11{,}712 = 3{,}022\ g/cm^2 \qquad (9.20)$$

When $\Delta\mu_{Na}°$ is converted to the action potential, E_a, the conversion factor is 1 kcal = 43.4 mV. Then:

$$t = 830\, c_{Na}''E_a \qquad (9.21)$$

For a typical muscle, mammalian or frog, c_{Na}'' is about 0.03 mol/kg water, and E_a is 120 mV. The calculated value of t is about 3,000 g/cm^2.

In the response of skeletal muscle to stimulation, the first signal is the action potential, followed at a measurable time interval by contraction. When the stimulation is tetanic, a constant maximal tension is developed. In frog muscle fibers, this is of the order of 3,000–3,500 g/cm^2, but in whole muscle it is only about one half that magnitude. From the above calculations, it is evident that maximal tension is proportional to the dielectric energy. Tension may thus be regarded as a colligative property, which depends on a change of state of the intracellular structured water. Electrolyte composition and electrical potentials are also determined by water structure.

In the literature, it is usually stated that the maximal isometric tension of frog muscle, stimulated by faradic current, is about 2,000 g/cm^2. The isometric tension of single isolated fibers is considerably higher [CASELLA, 1950]. The mean value for fibers from different muscles was 3,290 g/cm^2 at 25 °C. This is of the same order of magnitude as the values calculated for adult mammalian muscle from electrolyte composition (table XLIV). For frog muscle fibers, the maximal isometric tension is estimated from

Table XLIV. Maximal tension of isometric contraction

	Action potential $E_a$tmV[1]	c_{Na}''[2]	Calculated tension[3]	Observed tension[4] fiber	Observed tension[4] muscle
Frog, 20°C	120	0.030	3,020	3,290	2,500
					1,480
Frog 0°C					1,710
					1,085

1 Typical value, as given in table XLII.
2 Moles/kg water. KATZ [1896]; FENN [1935].
3 Grams/cm^2.
4 Values for individual fibers were independent of the nature of the muscle. Values 2,500 and 1,710 g/cm^2 represent the anterior tibialis. Values 1,480 and 1,085 g/cm^2 represent semitendinosus. The difference between the two muscles is anatomical.

the action potential and the sodium concentration as 3,020 g/cm^2. This agrees well with the value observed in isolated fibers, and is considerably higher than that observed for the different muscles: anterior tibialis and semitendinosus. Values for the tibialis were on the average 2,500 g/cm^2. This is of a smaller order of magnitude than the calculated values. Smaller orders of magnitude may possibly be explained on the basis of anatomical structure. Probably some of the fibers are not oriented parallel to the longitudinal axis of the muscle. In that case, tension should be exerted at the angle Θ with respect to the axis. Then:

$$t' = t \cos \Theta \qquad (9.22)$$

where t is calculated maximal tension, and t′ is the component that is exerted in the direction of the axis when the fiber is oriented at an angle of Θ degrees. The maximal value of t′ is obtained when Θ is zero.

When muscle is subjected to faradic stimulation, isometric contraction is accompanied by increased uptake of oxygen, output of carbon dioxide, and production of heat. According to the conception of concentration as given in the section on muscular work, the water in the stimulated state, either isotonic or isometric, becomes disordered and characterized by a high dielectric constant that may approach 80. In this state, many processes of anaerobic glycolysis and oxidation may be activated, along with processes of aerobic oxidation [Hill, 1944]. In the breakdown of glucose, the free energy change ΔG would be given by:

$$\Delta G = \Delta H - T\Delta S$$

In a rapid isotonic contraction, ΔH would approach zero, and the external work would be given by − TΔS. In an isometric contraction, the external work, − ΔG, becomes zero. Then:

$$\Delta H = T\Delta S$$

The system of contractile proteins, water and electrolytes can retain a state of constant low entropy by an increased rate of metabolic heat production. Then, in the breakdown of glucose to lactic acid:

$$\Delta G = -29.88 \text{ kcal}; \ \Delta H = -17.67 \text{ kcal}; \ T\Delta S = 12.22 \text{ kcal}$$

Therefore, in isometric contraction, the muscle loses free energy at a greater rate than enthalpy. Free energy is converted to heat rather than to work. The 'negative entropy' is used to maintain length and order in a system of high dielectric constant. The tension corresponds to that of isometric fibers in a medium of lowered configurational free energy of the intramuscular dispersion medium.

As will be shown in following chapters, the anaerobic breakdown of glucose implies a high respiratory quotient, loss of free energy and 'negative entropy', and a state of imbalance between ΔG and ΔH. Therefore, a prolonged state of isometric contraction would lead to changes of intracellular chemical morphology and of water and electrolyte balance. Recovery of the normal standard state requires a period in which energy and water balance is restored. This implies a period of high fat metabolism, lowered carbohydrate utilization, leading to a lowered respiratory quotient. In this period water returns to a state of low entropy, with altered solvent properties. The state of water, which varies with muscular tension and work in this way determines changes in cellular respiration and metabolism and changes in water and electrolyte balance.

The foregoing results for mammalian and frog muscle fibers may be compared with observed isometric tensions in invertebrate muscle. HOYLE and SMITH [1963] have studied electrical potentials and mechanical tension in the giant muscle fibers of a barnacle *Balinus nubilus* Darwin. The fibers were immersed in a 'barnacle Ringer's solution' of the following composition in mM/l: Na, 176; K, 8; Ca, 20; Mg, 12; Cl, 536; HCO, 10. Resting potentials of -74 to -90 mV were observed. In tetanic contraction of whole muscle maximum forces of 1,800 g/cm^2 were observed. This is of the same order of magnitude as the observed isometric tensions found in whole muscle of the frog, but is only about 60% of the tension of isolated frog fibers.

In single isolated fibers of the barnacle, tensions up to 35 g were observed; the cross sectional area was about 0.01 cm^2. The results correspond to a tension of about 3,500 g/cm^2. This is of the same order of magnitude as the observed and calculated tensions for frog muscle. It is possible that the dielectric energy $c_{Na}'' \Delta\mu_{Na}°$, is also of the same order of magnitude. However, in invertebrates such as the crab, *Carcinus maenas*, where c_{Na}'' is high (54 mM/kg water), the dielectric energy is also very high (table XLIII). This could lead to high isometric tensions in crustacean muscle. However, in this type of muscle, it is possible that the maximal theoretical tension is never developed. One reason for this is that all the individual fibers are not parallel to the longitudinal axis of the muscle. Another possible reason is that in tetanic contraction, the full dielectric energy is not developed.

Chemical Morphology and Dielectric Energy

At any period of the life span, normal chemical morphology depends on the well-ordered synthesis of *n* macromolecular substances, which leads to the embryological and ontogenetic development of each kind of cellular and extracellular structure. If the set of *n* macromolecular components in a certain kind of muscle fiber is denoted as $S_1, S_2 \ldots S_n$ at any given time, it is characterized by the composition of the macromolecular aggregates denoted by the individual concentrations $C_1, C_2 \ldots C_n$.

Chemical morphology also requires for its description a set of five ion concentrations: $(c_{Na})''$, $(c_K)''$, $(c_{Ca})''$, $(c_{Mg})''$ and $(c_{Cl})''$. Water is also an integral part of the aggregate. Unlike water in the fluids of the *milieu intérieur,* intracellular water is, to a great extent, structured rather than perfectly miscible. It is the dispersion medium for the soluble fractions of the five inorganic ions, as well as for bicarbonate, hydrogen ions and all of the soluble intracellular components. Thus the chemical morphology in the standard resting state is described by the properties of a system of at least $(n + n')$ substances, where *n* is the number of macromolecular components, and *n'* is the number of other substances necessary to determine chemical composition.

In the analyses of muscle, as given in tables VIII–XXIII, composition is described with reference to five inorganic ions, expressed as equivalents or moles per kg water. Although the colloidal composition is not given, it is implied by species and the type of muscle. Thus dielectric energy, electrical potentials and dielectric constant in subsequent calculations are assumed to be implied by a minimal number of $(n + 6)$ substances, which determine chemical morphology. Here *n* is treated as a constant that denotes the number of macromolecular components, and 6 refers to water and the set of five ions. In normal muscle in the processes of growth and development, the *n* macromolecular components are synthesized synchronously rather than independently. This is necessary at any point on the growth curve to produce a system with only one degree of freedom in the normal resting state. Then all the properties of the resting fiber are implied by the primary set of $(n + 8)$ parameters that includes the properties of normal temperature and pressure. These may be denoted as A, B, C... Then:

$$A, B, C \ldots \supset X, Y, Z \ldots$$

where the secondary properties, X, Y, Z... include permeability, metabolism, transport of oxygen, carbon dioxide and nutrients, and the electrical properties such as resting potential, conductance and impedance. These are fixed in the standard state by the standard values of A, B, C . . . This state is a function of growth and development of a system that is well-ordered and constrained with respect to nutrition and chemical morphology.

In a normally well-ordered physiological response, and in a well-controlled isotonic or isometric contraction, the resting set of secondary properties, X, Y, Z . . . passes over (with one degree of freedom) to a stimulated set, X′, Y′, Z′ . . . In this process, external work is determined by the change of dielectric energy. This is proportional to the change of standard chemical potential of sodium, $\Delta\mu_{Na}{}^{\circ}$, which is determined by a change of state of water and dielectric constant, D″. In this process, the free energy change, ΔG_c, is configurational, and is determined by $-T\Delta S_c$, where S_c is configurational entropy.

The change of configurational entropy is positive, work is positive, and ΔG_c is negative. Thus the work is produced by a very rapid change of order and disorder in the fiber rather than by intramuscular chemical reactions that involve the breaking or formation of chemical bonds. Normal chemical morphology of muscle fibers is constant in the resting state; the stimulated state is characterized by a change of dielectric properties such as dielectric energy, standard chemical potentials and dielectric constant, D″. In a rapid isotonic contraction these may change simultaneously, which no change of chemical composition. The change may be expressed as:

$$(\mu_i{}^{\circ}, D'', S_c, G_c)' \rightarrow (\mu_i{}^{\circ}, D'', S_c, G_c)''$$
$$(\mu_i^{\circ}, D'', S_c, G_c) \supset X, Y, Z \ldots$$

where X, Y, Z . . . is the well-ordered set of secondary properties. In a response with one degree of freedom, the independent variable may be taken as D″. If, as in a rapid isotonic contraction, this changes simultaneously with tension and state:

$$(D''_{A, B, C \ldots}) \supset X, Y, Z \ldots$$

and

$$(D''_{A, B, C \ldots})' \supset X', Y', Z' \ldots$$

The resting state may be regarded as the standard state of maximal stability, which can pass to a normally accessible state of excitation. In the stimulated state, chemical morphology of the fibers may change only with respect to length, tension, dielectric energy, entropy and free energy. All of these properties may be expressed in relation to D″ (or $\Delta\mu_{Na}{}^{\circ}$). In a system with one degree of freedom, it is possible for all the secondary functions to depend on one independent variable (D″), which is related to the physical properties (order or disorder) rather than to chemical composition or bond energies. In properties that involve irritability or work, entropy or disorder depend only on intracellular organization of the macromolecular components, rather than on changes in their chemical composition. Thus, when order, disorder and state are included in the concept, it can be said that all secondary physicochemical and physiological properties are implied by the chemical morphology of living cells and tissues.

Chapter X: Energy and Entropy

Energy means a capacity of performing work or of causing physicochemical change; entropy is a measure of the irreversible conversion or degradation of energy. Changes may occur under conditions of constant energy provided that there is an increase of entropy. Physicochemical systems may also change when the energy decreases at constant entropy. The physicochemical changes that occur when energy is transformed may be irreversible. Entropy is the thermodynamic measure of irreversibility. In studying irreversible changes and processes, the entropy changes and the energy changes are considered simultaneously rather than independently. All spontaneous processes in nature, when regarded in their totality, are irreversible. All are accompanied by increases of entropy in a closed or isolated system. Thermodynamics is the science which derives the physicochemical consequences of the energy and entropy principles, expressed quantitatively and related to measurable phenomena. In the following pages, these principles will be used to derive a number of fundamental equations which are applicable to certain important classes of physicochemical equilibrium. These derivations will illustrate the ways in which thermodynamic equilibrium depends on physical conditions such as temperature, pressure and the properties of matter in various states of aggregation.

General Criteria of Equilibrium

Given a fixed quantity of a pure chemical substance or mixture of substances isolated from its surroundings with respect to exchanges of heat or matter, this isolated quantity can be defined as a physicochemical system at constant energy. The system may or may not be in equilibrium. If not initially at equilibrium, the internal spontaneous processes will eventually tend toward a state of equilibrium. Stated mathematically:

$$(\delta S)_E \leqslant 0 \tag{10.1}$$

where S is the entropy and E is the energy. The symbols state: 'In any system at constant energy, E, an imaginary or virtual displacement from the state of equilibrium would imply a decrease of the entropy, S, and this decrease would approach zero in the limit.' Equation 10.1 implies as a corollary:

$$(\delta E)_S \geqslant 0 \tag{10.2}$$

At equilibrium in a system of constant entropy, the energy is at a minimum. Equation 10.1 and 10.2 are equivalent in the sense that one necessarily implies the other. This is shown in figure 1.

When entropy is increased from S_1 to S_2 at constant energy, E_1, the system reaches a state of maximal entropy at constant energy (equation 10.1). This point is represented by B (fig. 1). When the energy is increased to E_2, the corresponding increase of entropy is represented by a parallel line. In the higher state of energy, at a point corresponding to S_2, the system is not at equilibrium. A vertical line dropped to the point of B again represents a state of equilibrium at that point. Point B satisfies the condition of equation 10.2, which requires minimal energy, E_1, for systems at constant entropy, S_2.

Entropy is defined as a mathematical function which determines temperature equilibrium in a closed, isolated system of constant energy.

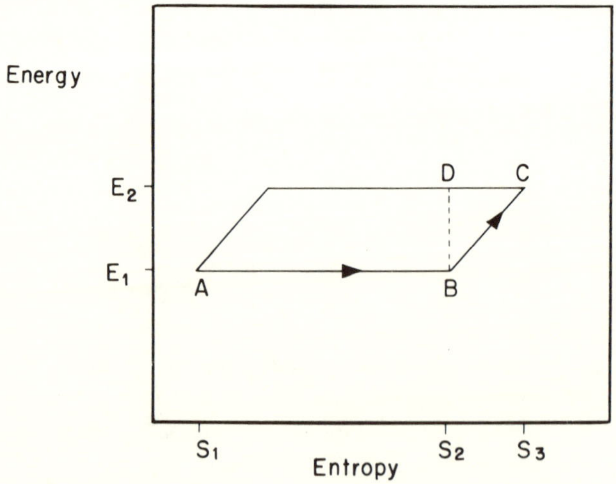

Fig. 1. Equilibrium conditions in a system of constant entropy.

According to equation 10.1, thermal equilibrium in such a system is determined by the condition that for any exchange of heat energy between two parts of the system, the change of entropy is zero. Let us imagine that q calories of heat energy are transferred from region A, where the absolute temperature is T_A, to region B at temperature T_B. Let the entropy change at A be expressed by $-q/T_A$, and the entropy change at B to be expressed as q/T_B. Then the net change of entropy, ΔS, is:

$$\Delta S = q/T_B - q/T_A$$

The condition for thermal equilibrium is that ΔS be zero, or that T_A be equal to T_B. The condition of spontaneous flow of heat from A to B is that ΔS be positive, or that T_A be higher than T_B. According to the above equation, there is a decrease of entropy in the region of A, and an increase of entropy in the region B, but the net change of entropy is positive. In any system containing regions A, B, C . . . Z, temperature equilibrium is attained when the temperature is uniform. This is the condition for maximal entropy at constant energy. In general, the entropy change may be expressed by the formula:

$$\Delta S = \frac{q}{T} \tag{10.3}$$

when it is understood that the heat q is added reversibly at the absolute temperature T.

Thermodynamic Functions

In addition to the internal energy; E, and the entropy, S, the following functions express the thermodynamic properties of pure chemical substances. All are defined in relation to the energy and entropy and to the absolute temperature, T, the pressure, P, and the volume, V. The molal enthalpy, H, is defined as:

$$H = E + PV \tag{10.4}$$

In general for pure substances, H, E and V are molal quantities; they refer to 1 mole of a chemical substance in a well-defined state at a certain temperature and pressure. The functions A and G are likewise molal quantitities. The work function, A, also known as the Helmholtz free energy, is defined as:

$$A = E - TS \tag{10.5}$$

where S is the molal entropy. The molal free energy (or Gibbs free energy) is defined by the relation:

$$G = E + PV - TS \tag{10.6}$$

From the above definitions, it follows that:

$$G = H - TS \tag{10.7}$$

and that:

$$G = A + PV \tag{10.8}$$

The relation of G and A to physicochemical equilibrium under various conditions will be considered. Energy is expressed in relation to an added increment of heat, dq, and an infinitesimal expenditure of external work, pdv. The entropy change, dS, in an infinitesimal reversible change can be related to dq:

$$dq = TdS \tag{10.9}$$

The value of dS over any closed cyclic path is zero; it depends only on the coordinates of the initial and final states. When the initial and final states are the same, the integral becomes zero. Thus dS has the properties of an exact differential, whereas dq is inexact. It denotes an increment of heat rather than the differential of a function of state. In considering the exchanges of heat energy and their relation to external work, it is necessary to eliminate q and w from the fundamental equations, and to retain as independent variables only those functions which yield exact differentials. Functions of this kind depend only on those conditions which determine the energy and the entropy: these quantities are determined by two of the three variables, P, V and T, which define the state. When entropy and volume are taken as the independent variables:

$$dE = TdS - PdV \tag{10.10}$$

where dS and dV are exact or perfect differentials. By applying the definition of G (equation 10.6), it is found that:

$$dG = -S\,dT + V\,dP \tag{10.11}$$

In equation 10.10 energy is expressed as a function of two independent variables, S and V. By defining the free energy according to equation 10.6, it is seen that temperature and pressure become the independent variables. Equations 10.10 and 10.11 are equivalent. This means that if there is a known relation between E, S and V for any chemical substance, then the relation between G, T and P is implicit in that relation. The converse proposition is also true. A known relation between G, T and P implies a relation between E, S and V, which can be derived from the known relation. This is proved in the following way. Let

$$E = f(S, V)$$

be a known relation between E, S and V. Then from equation 10.10:

$$\frac{\partial E}{\partial S} = T \tag{10.12}$$

and

$$\frac{\partial E}{\partial V} = -P \tag{10.13}$$

Temperature and pressure can be expressed as partial derivatives of E with respect to S and V, respectively. Free energy, G, is expressed in terms of these five variables; accordingly the values of S and V, which are sufficient to determine E are also sufficient to determine G. According to equation 10.11:

$$\frac{\partial G}{\partial T} = -S \tag{10.14}$$

and

$$\frac{\partial G}{\partial P} = V \tag{10.15}$$

The partial derivatives of G with respect to T and P therefore determine S and V. This suffices to yield a relation between E, S and V, which can be derived from any known relation between G, T and P. Consequently either relation can be derived from the other. It can be shown in this way that equations among the following four sets of variables are equivalent in this sense. A known relation between E, S and V is equivalent to one between H, S and P, and so forth:

(E, S, V); (H, S, P); (A, T, V); (G, T, P).

Equilibrium at Constant Temperature

It is a condition of thermal equilibrium in a physicochemical system that themperature be constant in every region (equation 10.3). Treatment of the equilibrium conditions at constant temperature requires the introduction of additional thermodynamic functions, namely the free energy, G, and the work function, A. A familiar example of an equilibrium at constant temperature is that between water and water vapor (or steam) at one atmosphere pressure and 100 °C. The molal heat of vaporizetion of water is about 9,700 cal. This is expressed as an increase of enthalpy and is symbolized as $\Delta H_{373.1}$. The sign is positive because heat is required to form steam from water. Other thermal and mechanical changes may be expressed from the known properties of water and steam. The entropy change, $\Delta S_{373.1}$, as calculated from ΔH and the absolute temperature, is about 26.0 entropy units (cal/mol/deg.). In general, at equilibrium:

$$\Delta S = \frac{\Delta H}{T} \qquad (10.16)$$

Since the pressure is 1 atm, the volume change, ΔV, is given by:

$$\Delta V = 22.4 \times \frac{373.1}{273.1} = 30.3\ \mathrm{l/mol}$$

The molal volume of liquid water is very small compared to that of water vapor, so that ΔV can be estimated from the volume of the water by means of the ideal gas law. The expansion is carried out at constant pressure so that the work, W, is:

$$\begin{aligned} W &= P\Delta V \\ &= 30.3\ \mathrm{l\ atm} = 745\ \mathrm{cal} \end{aligned}$$

In any process at constant temperature, the expression for ΔA is:

$$\Delta A = \Delta E - T\Delta S \qquad (10.17)$$

The two terms of the right hand member are:

$$\Delta E = q - w; \quad T\Delta S = q$$

Therefore:

$$\Delta A = -W = -745\ \mathrm{cal} \qquad (10.18)$$

The work of expanison, W, in any process at constant temperature is measured by − ΔA. Hence ΔA is a force function at constant temperature just as ΔE is a force function at constant entropy. ΔE measures the maximum work yielded by the physicochemical change at constant entropy, while ΔA measures the maximal work at constant temperature. The criteria of equilibrium under the respective conditions are:

$$(\delta E)_S \geqslant 0$$

and

$$(\delta A)_T \geqslant 0$$

We may now compare the value of ΔA with that of ΔG for the vaporization of water at 100 °C and 1 atm pressure:

$$\Delta G = \Delta E + P\Delta V - T\Delta S \quad (10.19)$$

$$= \Delta H - T\Delta S \quad (10.20)$$

$$= \Delta A + P\Delta V \quad (10.21)$$

Therefore $\Delta G = 0$ for the vaporization of water at constant temperature and pressure. This condition is also true for any other change of state under equilibrium conditions in an open system at constant temperature and pressure. The criterion for equilibrium as expressed by ΔG is always:

$$\Delta G = 0$$

The condition for a minimal value of G is expressed by:

$$(\delta G)_{T,P} \geqslant 0 \quad (10.22)$$

Thus δG is a force function at constant temperature and pressure, just as δE is a force function at constant entropy. Equation 10.22 applies to other changes of state of water, as for example, freezing point equilibrium or the sublimation of ice. It also applies to the equilibrium between cells, tissues and the *milieu intérieur* of living organisms. In all changes of state at equilibrium temperature and pressure, the free energy is minimal, and the free energy change of transporting water between the two phases is zero.

Differential Equations

'The point of mathematics is that in it we have always got rid of the particular instance, and even of any particular sorts of entities' [WHITEHEAD, 1925]. In theoretical physics, for example, the fundamental laws of dynamics, electrostatics, optics and acoustics may be expressed by equations of almost identical form. Similarly, the equations or sets of equations describing heterogeneous equilibrium are of the same form when they are applied to living cells and tissues as when they are applied to any other kind of phase rule equilibrium that involves only non-biological substances [JOSEPH,

1971a, b]. In the examples of heterogeneous equilibrium that were given in chapters I–IX, chemical morphology and state were shown to conform to fundamental thermodynamic functions such as configurational free energy and entropy.

The simplest kind of physicochemical system is that which consists of a pure chemical substance in the solid, liquid or vapor state. For one mole of the substance, the physicochemical state is determined by any two of the three variables, temperature, pressure and volume. Often it is most convenient to define the state by the temperature and pressure. If two or more chemical substances are present, it is also necessary to specify the composition of the mixture in addition to the temperature and pressure. Physicochemical state therefore always depends on at least two independent variables. When the material composition is variable, as it is in biological systems, at least (n +8) quantities are required to describe composition and state. In dealing with the properties of systems of this kind, it is necessary to employ differential equations with several independent variables. This requires the application of the methods of partial differential equations.

In general, if a quantity, u, depends on a set of several independent variables, x, y, z . . ., we assume that it can be expressed within certain limits as a continuous single valued function:

$$u = f(x, y, z \ldots) \tag{10.23}$$

For such a function, u, an infinitesimal change, du, is related to infinitesimal changes of the independent variables.

$$du = X\,dx + Y\,dy + Z\,dz \tag{10.24}$$

where the coefficients, X, Y and Z depend on the relation between u, x, y and z. The function u is expressed in this form as the dependent variable, which is a secondary property. The coefficients X, Y and Z are defined as the partial differential coefficients of u with respect to x, y and z.

$$X = \frac{\partial u}{\partial x}; \quad Y = \frac{\partial u}{\partial y}; \quad Z = \frac{\partial u}{\partial z} \tag{10.25}$$

The same value for the second derivative is obtained if the coefficient

$$\frac{\partial u}{\partial y}$$

is differentiated with respect to x. This is a general rule for second partial differential coefficients with respect to two independent variables. It follows from fundamental principles that:

$$\frac{\partial T}{\partial V} = -\frac{\partial P}{\partial S} = \frac{\partial^2 E}{\partial V\,\partial S} = \frac{\partial^2 E}{\partial S\,\partial V} \tag{10.26}$$

Equation 10.26 is one of four similar thermodynamic relations between temperature, pressure, volume and entropy. These are known as the Maxwell relations. They are derived from the following:

$$dE = TdS - PdV \tag{10.10}$$

$$dH = TdS + VdP \tag{10.27}$$

$$dA = -SdT - PdV \tag{10.28}$$

$$dG = -SdT + VdP \tag{10.29}$$

It has been shown that a known relation between E, S and V is equivalent to one between G, T and P. It is also equivalent to relations between H, S and P, and between A, T and V. The four fundamental differential equations are therefore equivalent. By forming the second differential coefficients with respect to the two independent variables, the following four Maxwell relations are obtained:

$$\left(\frac{\partial T}{\partial V}\right)_S = -\left(\frac{\partial P}{\partial S}\right)_V \tag{10.30}$$

$$\left(\frac{\partial T}{\partial P}\right)_S = \left(\frac{\partial V}{\partial S}\right)_P \tag{10.31}$$

$$\left(\frac{\partial S}{\partial V}\right)_T = \left(\frac{\partial P}{\partial T}\right)_V \tag{10.32}$$

$$\left(\frac{\partial S}{\partial P}\right)_T = -\left(\frac{\partial V}{\partial T}\right)_P \tag{10.33}$$

It is evident from the foregoing that entropy, temperature, pressure and volume are connected by an intricate ramification of relationships. It is not possible to change one of the independent variables without simultaneously changing all the dependent variables and their first and second derivatives. It is quite impossible to comprehend the multiplicity of these ramifications without further attention directed to the methods of partial differential equations. Visual representation of correlated sets of dependent and independent variables is facilitated by the use of Cartesian and d'Ocagne nomograms (chapter XIII).

Thermodynamic Functions as Line Integrals

If a material particle is in field of force which depends on the coordinates x, y and z, there are three directional components of the energy and of the work necessary to change the position of the particle in the field. If U is the potential energy at a point, the infinitesimal change of energy, dU, is related to dx by:

$$dU = X\,dx$$

where X is the component of force parallel to the x axis. If there are simultaneous changes of y and z, then:

$$dU = X\,dx + Y\,dy + Z\,dz$$

The components of force, X, Y and Z may be expressed as partial derivatives of U with respect to x, y and z. Then:

$$dU = \frac{\partial U}{\partial x} dx + \frac{\partial U}{\partial y} dy + \frac{\partial U}{\partial z} dz$$

The change of potential energy between any two points (x_1, y_1, z_1) and (x_2, y_2, z_2) is given by:

$$U_2 - U_1 = \int (X\,dx + Y\,dy + Z\,dz)$$

If the potential energy is a function only of x, y and z, then its first derivatives are also functions of the coordinates. The integral is an example of a *line integral,* the value of which depends only on the initial and final sets of coordinates. The value is independent of the path, C, over which the line integral is extended. The necessary and sufficient condition of this independence is that U be a continuous single valued function of x, y and z. The differentials dx, dy and dz are then considered to be exact differentials. In the expression:

$$dq = dE + P\,dV$$

the fact is expressed that an infinitesimal quantity of heat, dq, when added to a gas, increases the internal energy by an infinitesimal amount, dE, at the same time that external work, P dV, is performed. If V and P are taken as the two independent variables,

$$dq = X\,dV + Y\,dP$$

where

$$X = \frac{\partial E}{\partial V}$$

and

$$Y = \frac{\partial E}{\partial P}$$

The heat increment dq is not an exact differential. Therefore the expression (X dV + Y dP) is not an exact differential. It cannot yield a line integral which depends only on the coordinates V and P. When the integral

$$\int (X\,dV + Y\,dP)$$

is extended over a closed path, C, it represents the heat absorbed in a cyclic process. It has been shown that this is zero only in the limiting case where there is no change of temperature (isothermal, reversible cycle). This signifies that the heat, q, depends on the path C rather than on the independent variables, P and V. If an integrating factor, β, can be found which has the property that the integral

$$\int \beta (X\,dV + Y\,dP)$$

depends on the initial and final coordinates rather than on the path, then β (X dV + Y dP) would have the properties of an exact differential. In particular, the value of the integral extended over any closed path would be zero. The absolute temperature, T, which is a function of P and V yields through its reciprocal the required integrating factor. Thus the value of the integral

$$\int \frac{X\,dV + Y\,dP}{T}$$

is zero over any closed path. Therefore, dq/T has the properties of an exact differential. This differential has been defined as dS, where S is the entropy. Accordingly, the change of entropy, ΔS, may be expressed as a line integral between any two points; entropy is therefore a function only of thermodynamic state rather than of path, process or mechanism. Accordingly:

$$\Delta S = \int \frac{dq}{T}$$

depends only on the change of state that is expressed by the limits of the definite integral. Therefore, in an isothermal reversible process integrated over a closed cycle, ΔS is zero (Carnot's principle). Over such a closed path, ΔG and ΔH are also zero. Therefore no energy or work is required to maintain an invariant state of water and electrolyte balance in a heterogeneous biological system [JOSEPH, 1966a]. This contradicts the fundamental postulate of the 'sodium pump', which depends on the idea of mechanisms of 'ion transport' that require continuous supplies of bond energy. Such a continuous expenditure of bond energy to maintain state is therefore incompatible with both the first and second laws of thermodynamics [JOSEPH, 1966a, 1971a, b].

By defining dS as an exact differential, the inexact differentials, dq and dw, are eliminated. Then:

$$T\,dS = dE + P\,dV$$

where dE and P dV are also exact differentials. The other functions, H, A and G are defined in terms of the quantities, E, T, S, P and V, all of which yield exact differentials, and which therefore depend only on thermodynamic state rather than on the path between two states. Accordingly, ΔH, ΔA and ΔG depend only on the initial and final states of any physical or chemical process. For example, ΔG is expressed as a function of temperature and pressure, and for a pure chemical substance undergoing a change of the two independent variables, would be determined by the line integral:

$$\int (-S\,dT + V\,dP)$$

The value of the integral is independent of the path C.

The definition of entropy and free energy changes as line integrals is based on the application of the Pfaffian differential equations of pure mathematics. Thermodynamics than becomes a branch of pure mathematics, based on the introduction of T as the integrating denominator of dq. The second law is founded on the principle that there are inaccessible states in the neighborhood of a given accessible state [CARATHEODORY, 1909; BORN, 1948]. Biological stability may be thought to depend on such constraints [JOSEPH, 1971a].

Biological Applications of Line Integrals

The physical ideas leading to the expression of entropy as a measure of irreversible phenomena are borrowed from the theory of heat, and are not generalized to include all physicochemical phenomena, including those characteristic of biological systems. The fundamental concepts of classical thermodynamics were derived from engineering, as applied by physicists and engineers from Carnot to Kelvin. The final form of the second law receives expression by means of line integrals and exact differentials, and becomes free of ideas derived from irreversible work, as performed by heat engines.

In mammalian cells and tissues, fluctuations of temperature are very small. Therefore, transport of the inorganic physiological ions such as sodium, potassium and chloride is isothermal and reversible.

These ions are of environmental origin, and do not enter into any irreversible biological processes that involve chemical change. Any transport of these ions between cells and extracellular fluids is isothermal and reversible. The entropy change of such a cyclical process is given by the line integral

$$\int \frac{dq}{T}$$

Over a closed isothermal cycle, the value of $T\Delta S$ is zero. This is a condition for invariance in the system. Another condition is that ΔG is also zero. Thus transport of sodium, potassium and chloride ions in the standard state of all cells and tissues is isothermal and reversible. It requires no expenditure of work or other forms of energy. Actual expenditure of work or free energy would be measured as a negative value of ΔG or an increase of entropy. This is inconsonant with the principles of homeostasis or invariant steady states [JOSEPH, 1971a, b]. Cells are not required to spend metabolic energy or chemical bond energy in order to maintain an invariant state of constant entropy and energy. Metabolic heat produced by changes of chemical bond energy is transported to the external environment, as a condition of constant temperature. The transport is quantitative and irreversible, and conforms to the first law of thermodynamics. Invariant intracellular conditions imply second law restrictions on thermodynamic state. This state is maintained by the constant input of nutrients and metabolites, occurring at the same rate as intracellular respiration. Homeostasis may be defined as a self-identity, which is maintained as a standard state that shows minimal changes of composition, temperature, physicochemical and physiological properties.

A state of perfect electrolyte balance thus implies a state of perfect energy balance, in which all transport processes are constant. It can therefore be said that constant transport is maintained by a constant standard state of composition and respiration, attended by the simultaneous operation of both reversible and irreversible (or exothermic) processes. The organism thus behaves as an open system with respect to nutrients, respiratory gases and heat. However, it retains constant chemical morphology, invariant internal and external boundaries, and fixed conditions of state. The constant state of cells and tissues, including the *milieu intérieur*, requires a set of thermodynamic conditions of invariance and constraint, which place restrictions on the 'open system'. The principles which determine energy balance as well as water and electrolyte balance in biological systems are based on GIBBS' theory of 'equilibrium of heterogeneous substances'. This has been shown to be applicable to living systems of cells and tissues, as well as to the very great number of non-living systems to which it has been applied in physical chemistry.

Chapter XI: Heterogeneous Equilibrium

The prevailing state of matter on the land surface of the Earth is heterogeneous, impure and non-ideal. To a large extent, the homogeneous pure substance is the result of man's intervention in nature. The ideal state of matter is a product of his imagination. In his attempts to create order out of disorder, he produces all kinds of manufactured products, and produces many synthetic derivatives from natural products.

One of the most complicated of all physicochemical systems is the human body, in which innumerable chemical substances of all kinds are inextricably mixed in a very large variety of cells, tissues and organs. The system consists of a very large number of immiscible phases, which are heterogeneous at all levels, down to submicroscopic structures which can be visualized only with the electron microscope. Liquids such as gastric juice, saliva, bile, lymph, blood plasma and other body fluids coexist within the same organism, as do muscles, nerve, bone, cartilage and skin. All have their characteristic composition of water, minerals, proteins and other macromolecular substances.

All anatomical, morphological and physiological functions are related to the distribution of matter in heterogeneous physiological systems. This distribution follows the general laws of thermodynamics, described by GIBBS' 'equilibrium of heterogeneous substances'. These laws are formulated in the phase rule, which is based on the first and second laws of thermodynamics. Phase rule is far more general than the simplified special case of 'membrane equilibrium', described by DONNAN for systems in which water exists only in miscible liquid phases. These systems show the property of perfect mixing. Therefore they characterize only certain limiting classes of phases within the body. All others are immiscible; their chemical composition and physicochemical properties depend on negative configurational entropy that is characteristic of biological structures that exist as solid or semi-solid macromolecular aggregates rather than as homogeneous, miscible liquids.

'A solution is defined as a homogeneous mixture which can pass through variations of composition within the limits that define its existence' (chapter II). Therefore, a large number of biological structures cannot be described as solutions. They are heterogeneous, rather than homogeneous mixtures, and in the standard state are of invariant composition; they do not pass through variations of composition extended over any wide range. They are restricted and constrained systems with a small number of degrees of freedom rather than true solutions. With the same number of freely diffusible components, true solutions would have a very large number of degrees of freedom, and would show very widely variable composition and state. The thermodynamic conditions of invariance that are applicable to heterogeneous biological systems are based on the general laws of heterogeneous equilibrium and can be applied to biological systems by means of the phase rule [JOSEPH, 1971a, b].

Fundamental Equations

The phase rule is based on derivations for equilibrium in heterogeneous systems. In any phase:

$$dG = -S\,dT + V\,dP + \Sigma\mu_1\,dm_1 \quad (11.1)$$

Free energy, G, is a function of (n + 2) independent variables, where *n* is the number of components, and 2 stands for temperature and pressure, the external variables which are factors in the states of equilibrium. In the following treatment, these will be considered to be the only significant external variables. Gravitational, capillary and electrical fields of force are not considered in the fundamental derivations, but are applied to special problems.

For a homogeneous phase of one component:

$$dG = -S\,dT + V\,dP \quad (11.2)$$

This represents the molal free energy of a pure component in one of its possible states of aggregation: liquid, solid or gas. The free energy is expressed as a function of temperature and pressure. Variations of total mass are not considered. In this fundamental equation, five variables are included, of which two, the temperature and the pressure, are the independent variables. One, the free energy, is the dependent variable. The other two, entropy and volume, may be expressed as the partial derivatives of free energy with respect to temperature and pressure. Entropy is a conjugate variable of temperature, and volume is a conjugate variable of pressure. In any specific case, the relation between G, T and P is determined by the nature of the pure component in the given state of aggregation. It is equivalent to an equation of state. By means of this relation, the conjugate variables, S and V, may also be expressed as functions of the two independent variables, T and P. They are obtained as partial differential coefficients of G with respect to T and P:

$$\frac{\partial G}{\partial T} = - S = f_1 (T, P) \tag{11.3}$$

and

$$\frac{\partial G}{\partial P} = V = f_2 (T, P) \tag{11.4}$$

The equation of state is expressed as:

$$G = f (T, P) \tag{11.5}$$

These three equations express general relations among the five variables. Since only two variables, T and P, are independent, all five variables may be expressed in terms of these two. The partial derivative of G with respect to P yields an expression for V as a function of T and P. For ideal gases, this assumes the form of the ideal equation of state. For real gases, the deviations appear in the equation of state (equation 11.5), and in the partial derivatives.

If, instead of a pure substance, a mixture were considered in one of its possible states of aggregation, or phases, the free energy may be expressed as a function of T, P and the masses of the *n* components. Then:

$$G = f (T, P, m_1 \ldots m_n) \qquad (11.6)^1$$

1 In equation 11.6 and throughout chapter XI, *n* is used to denote the total number of soluble components in a homogeneous solvent or dispersion medium. These include the two binary electrolytes, NaCl and KCl, and the two ternary electrolytes, $CaCl_2$ and $MgCl_2$. It also includes all soluble components that contribute to the free energy of the phase. Thus *n* is used in a different sense from that of preceding chapters where it has been employed to donote the set of *macromolecular* components in a given cell or tissue. In the preceding sections, the system was characterized by a univariant set of (n + 6) substances, where the number 6 referred to water and the five inorganic ions. In chapter XI, if the set of colloidal components were to be included, it would be possible to denote the number of such substances as n′. Then the system could be considered to consist of (n′ + n) components, where *n* is the number of substances that may be characterized by chemical potentials.

The free energy is expressed as a function of (n + 2) variables, as in equation 11.1. The total number of variables is (2n + 5). These include G, S, T, V and P, in addition to *n* masses and *n* chemical potentials. If G is a known function of the (n + 2) independent variables, then S, V and the *n* chemical potentials may be defined as the first partial derivatives. There are thus (n + 2) first derivatives of free energy with respect to the independent variables. These derivatives are the conjugate variables of the system: one for each independent variable. Given the relation between G and the independent variables (the equation of state), there is a total of (n + 3) determinate relations among the (2n + 5) variables that characterize the system. For any phase, these relations are sufficient to determine completely the thermodynamic properties. The existence of (n + 3) equations among the (2n + 5) variables is sufficient, because (n + 2) variables are independent. These determine the total mass and composition of the phase and the conditions of temperature and pressure. They also determine the free energy, entropy, volume, and the chemical potentials of each of the components. The (n + 3) dependent and conjugate variables are functions of the (n + 2) independent variables. Since a known equation of the form of 11.6 is sufficient to determine G, S and V as functions of T and P, it is also sufficient to determine E, H and A. These functions are related to the formulae:

$$E = G - PV + TS$$

$$H = G + TS$$

and

$$A = G - PV$$

Accordingly, a known relation between free energy, temperature and pressure is sufficient to determine E, H and A in relation to the same variables. Similarly, a known relation of the form of 11.6 is equivalent to each of the following:

$$E = f_e(S, V, m_1 \ldots m_n)$$

$$H = f_h(S, P, m_1 \ldots m_n)$$

$$A = f_a(T, V, m_1 \ldots m_n)$$

Thus, E, H and A may also be expressed as functions of (n + 2) independent variables, and each expression is equivalent to the relation between G, T, P and the masses of the *n* components. Each of these four possible equations expresses a different set of conditions of constraint in the system with respect to temperature, pressure, entropy and volume. There is a fifth fundamental equation which is equivalent to the other four. It has been shown in earlier sections that:

$$- S\,dT + V\,dP = \Sigma m_1\, d\mu_1 \tag{11.7}$$

In this equation, the independent variables are temperature, pressure and the chemical potentials of the components. The masses of the *n* components are expressed as conjugate variables. Here, all the independent variables are *intensive* (zero order) rather than *extensive* (first order) quantities. They are independent of total mass or volume, and depend only on composition and state. There are therefore (2n + 4) variables, of which only (n + 1) are independent. Volume and total mass are not included in the set of independent variables. It is easily seen that the chemical potentials cannot all be varied independently. For example, at constant temperature and pressure, for a one component system, μ_1 would be constant. For a two component system:

$$m_1\, d\mu_1 + m_2\, d\mu_2 = 0$$

Therefore, only one potential may be varied independently. In general, only (n − 1) potentials may be independent variables. This corresponds to the fact that in a system of *n* components, the *n* chemical potentials are determined, not by the *n* masses, but by the (n − 1) ratios of all other masses to that of the solvent. Among the (2n + 2) variables:

$$f(T, P, \mu_1 \ldots \mu_n) = 0 \tag{11.8}$$

then S, V, $m_1 \ldots m_n$ may be expressed as functions of the same variables. Together with equation 7.8, this yields a total of (n + 3) relations among the (2n + 4) variables of equation 11.7. This is sufficient to characterize the system, because (n + 1) of the variables are independent. A relation of the form of equation 7.8 is therefore equivalent to the other four fundamental equations. It differs from the others in that the independent variables are all zero order quantities. Thus in the living mammalian organism, zero order functions would be independent of body weight. On the other hand, a first order function such as G, H or A tends to be propertional to total body mass.

An equation of the form of 11.8 is related to the Gibbs-Duhem equation (GIBBS' equation 97). This may be derived from Euler's theorem, as applied to zero order homogeneous functions.

Gibbs' Definition of Phase

'In considering the different homogeneous bodies which can be formed out of any set of component substances, it will be convenient to have a term which shall refer solely to the composition and thermodynamic state of any such body without regard to its quantity or form. We may call such bodies as differ in composition and thermodynamic state different *phases* of the matter considered, regarding all bodies which differ only in quantity and form as different examples of the same phase' [GIBBS, 1928].

According to this definition, different phases of a heterogeneous system always differ in composition; uniform composition, or perfect mixing occurs only in homogeneous solutions. In biological systems, such uniformity occurs only in blood plasma or other liquid phases of the *milieu intérieur.* Accordingly, blood plasma and other body fluids are of similar electrolyte composition, resembling mammalian Ringer solution both in ion concentrations and in the freezing point and chemical potential of water. On the other hand, the distribution of electrolytes and water in the solid or semi-solid aggregates of somatic cells and tissues is related to the thermodynamic properties of each kind of component in each kind of phase. Standard chemical potentials are related to solubilities. Thus the permanently stable and enduring 'concentration gradients' of any biological structured system are characteristic of the configurational free energy and entropy of the system. Contrary to the postulates of the theory of 'active transport', the ion distributions require no respiratory or metabolic energy for their maintenance. Since the organism behaves as a nearly perfect calorimeter according to the first and second laws of thermodynamics, all the repiratory and metabolic heat is transported quantitatively and irreversibly to the external environment. The efficiency of ion transport mechanisms is therefore zero [JOSEPH, 1966a, 1971a, b].

In the presence of Ringer solution or body fluids that contain the natural concentrations of the inorganic ions, cells and tissues are always heterogeneous. Therefore, they cannot be expected to conform to the laws of perfect solutions, as would be the case if the distribution of ions were determined by the classical Donnan equilibrium [DONNAN, 1911]. GIBBS' theory of 'equilibrium of Heterogeneous substances', which implies the phase rule, is far more general than Donnan's theory of 'membrane equilibrium' in which each of the two solutions is homogeneous, with the property of perfect mixing.

According to MACALLUM [1910, 1926], the primitive environment of paleozoic ages had the properties of an early form of sea water. Then the condition for the genesis of living cells required the formation of biological phases that contained the same sets of electrolytes and ions that were present in the surrounding fluid. Thus the property of heterogeneous structure was necessary to differentiate the intracellular sets of water and electrolytes from the extracellular set. It is also required for phylogenetic and ontogenetic development. Biological systems develop with the conservation of reversibility. This depends on the preservation of the characteristic thermodynamic states of intracellular and extracellular aggregates and the endurance of an invariant *milieu intérieur* [BERNARD, 1878]. Cells and tissues are then free to develop, but only under thermodynamic conditions of invariance and constraint. Transport of ions such as sodium and potassium is not unconstrained, but is determined by null values of the thermodynamic line integrals for ionic diffusion, necessary to maintain invariance and reversibility (chapter X). In growth, aging and phylogenetic development, these conditions are determined by the phase rule, which is necessary to relate chemical composition, thermodynamic state, and physicochemical properties to the developing morphology. In addition to the thermodynamic sets of controls, growth, development and evolution depend on the properties of the fundamental units of heredity, the genes and chromosomes.

The Chemical Potential

In physicochemical systems that contain more than one chemical substance, the free energy and other thermodynamic functions depend on the composition. If we imagine the free energy, G, to be related to the masses, m_1, m_2, $m_3 \ldots m_n$, of a system that contains *n* components, the differential equation is of the form:

$$dG = -S\,dT + V\,dP + \Sigma \frac{\partial G}{\partial m_1} dm_1 \qquad (11.9a)$$

or

$$dG = -S\,dT + V\,dP + \Sigma\mu_1\,dm_1 \qquad (11.9b)$$

The *chemical potential* of any component is thus defined as:

$$\mu_1 = \frac{\partial G}{\partial m_1} \qquad (11.10)$$

GIBBS' definition of potential was given as follows: 'If to any homogeneous mass we suppose an infinitesimal quantity of any substance to be added, the mass remaining homogeneous, and its entropy and volume unchanged, the increase of the energy of the mass divided by the quantity of the substance added is the *potential* for that substance in the mass considered.' Relating the potential, μ, to the thermodynamic functions, E, H, A and G, GIBBS found the following four equivalent relationships:

$$\mu_1 = \left(\frac{dE}{dm_1}\right)_{S,V} = \left(\frac{dA}{dm_1}\right)_{T,V} = \left(\frac{dH}{dm_1}\right)_{S,P} = \left(\frac{dG}{dm_1}\right)_{T,P} \tag{11.11}$$

Depending on the conditions of constraint for a given system, μ may be defined in any one of four ways. When a system is constrained by conditions of constant entropy and volume, it is *closed.* These conditions isolate the system with respect to exchanges of heat with the surroundings and also with respect to performance of mechanical work. In the other limiting case, that of constant temperature and pressure, the system is free to exchange heat and to perform external mechanical work, since it is unconstrained with respect to entropy and volume. The system is then open, and constrained only with respect to temperature and pressure, which unlike entropy and volume, are intensive variables, or zero-order functions. The general conditions for phase rule equilibrium usually apply to open systems, in which state is a function of temperature, pressure, and chemical composition rather than of mass or volume. To apply the phase rule only to closed systems is to restrict the number of degrees of freedom to zero [JOSEPH, 1971b]. Thus it is incorrect to assume that phase rule or thermodynamics are inapplicable to biological open systems. Living organisms are always restricted with respect to permissible variations of internal and external environments [BERNARD, 1878].

Euler's Theorem

If u is homogeneous and has continuous first partial derivatives, then

$$x\frac{\partial u}{\partial x} + y\frac{\partial u}{\partial y} + z\frac{\partial u}{\partial z} = nu \tag{11.12}$$

where *n* is the order of the function. This is a statement of Euler's theorem.

When u is a first order function such as volume, free energy, enthalpy or entropy, it yields partial molal quantities which satisfy equations of the following form:

$$G = \mu_1 m_1 \ldots + \mu_n m_n \tag{11.13}$$

Entropy, which is homogeneous and of the first order with respect to the masses of the components, conforms to the equation:

$$S = \bar{S}_1 m_1 \ldots + \bar{S}_n m_n \tag{11.14}$$

where $\bar{S}$ is the partial molal entropy.

Whereas G, V and S are of the first order with respect to the masses of the components, their first derivatives are of zero order. The first order functions are extensive quantities, and the zero order functions are intensive. They are interrelated by equations of the form 11.13 and 11.14. At constant temperature and pressure, the free energy, G, is a continuous function of $m_1 \ldots m_n$, the masses of the *n* components. The first partial derivative of G with respect to the mass of any component is likewise a function of the same independent variables. Thus:

$$G = f_o\,(T, P, m_1 \ldots m_n)$$

$$\mu_1 = f_1\,(T, P, m_1 \ldots m_n)$$

$$\cdots\cdots\cdots\cdots\cdots\cdots$$

$$\mu_n = f_n\,(T, P, m_1 \ldots m_n)$$

The chemical potentials are obtained by partial differentiation of G with respect to each of the components. The second partial differential coefficients are obtained by differentiating the chemical potentials with respect to $m_1 \ldots m_n$.

$$\frac{\partial^2 G}{\partial m_1\,\partial m_2} = \frac{\partial^2 G}{\partial m_2\,\partial m_1} \tag{11.16}$$

$$= \frac{\partial \mu_1}{\partial m_2} = \frac{\partial \mu_2}{\partial m_1}$$

The derivation of these relations is based on the identity of the two second partial derivatives. The method has been applied previously in deriving the four Maxwell relations. The same method is applied to all the other second derivatives. Then it is evident that:

$$m_1 \frac{\partial \mu_1}{\partial m_1} + m_2 \frac{\partial \mu_2}{\partial m_1} + m_3 \frac{\partial \mu_3}{\partial m_1} = 0$$

In this expression, m is the only independent variable, and the partial derivatives can be replaced by $d\mu_1$, $d\mu_2$ and $d\mu_3$. Therefore:

$$m_1\, d\mu_1 + m_2\, d\mu_2 + m_3\, d\mu_3 = 0$$

or

$$\Sigma m_1\, d\mu_1 = 0$$

an equation which is equivalent to Euler's theorem applied to several components. The equation is known as the Gibbs-Duhem equation. As stated here, it is applicable to a system at constant temperature and pressure. In the form of equation 11.7, it is applicable to systems in which the temperature and pressure are variable. By integration at constant T and P, equation 11.13 is obtained. It is an application of Euler's theorem to a homogeneous function of the first order. Free energy, enthalpy, entropy, the work function and volume are such functions, whereas the chemical potentials and other partial molal quantities are expressed as homogeneous functions of zero order. Euler's theorem is applicable not only to the free energy and chemical potentials, but also to all the other homogeneous functions.

In equation 11.10, the differential coefficient of μ_1 with respect to m_2 may be denoted as a_{12}, the other derivative being denoted as a_{21}. The same method is applied to all the other second derivatives. Then it is evident that:

$$a_{12} = a_{21}$$

$$a_{13} = a_{31} \ldots a_{23} = a_{32}$$

$$a_{1n} \neq a_{n1}\, a_{2n} = a_{n2} \ldots a_{(n-1)n} = a_{n(n-1)}$$

The total number of such coefficients is therefore n(n − 1). But each coefficient a_{12} is equal to another, a_{21}, through the second partial differential coefficients. This number increases approximately as the square of the number of components. It is evident that the coexistence of components in a physicochemical system implies reciprocal interactions between all possible pairs of components, and that the number of these interactions is approximately proportional to n^2. In general, a change in the quantity of any component leads to changes in the chemical potentials of all the others. It is evident from the foregoing that none of the *n* chemical potentials of a set of *n* components can be varied independently. Each member of the set varies simultaneously. Therefore, protoplasm behaves as a unit with reference to any change of state. The same is true of the ground substance of connective tissue. Such a system is characterized as univariant with respect to any change of state that involves only one degree of freedom. It is therefore impossible, by adding a single chemical substance to such a system to alter a given chemical potential in a specific sense. Any such variation would involve the response of all *n* chemical potentials as a well-ordered unit.

Phase Rule

A physicochemical system exists in two or more phases when any chemical substance which can pass freely between any of the phases can do so without change of free energy or chemical potential. Then the conditions of stability and reversibility require for all such substances that:

$$\mu_1' = \mu_1'' \ldots = \mu_1^{p}$$

$$\mu_c' = \mu_c'' \ldots = \mu_c^{p} \qquad (11.17)$$

This set of equations describes the condition of thermodynamic stability in a system of *c* components that are freely distributed among *p* phases. The phases may be described as 'different homogeneous bodies which can be formed out of the same set of component substances' [GIBBS, 1928].

In biological systems the set of equations 11.17 may be applied to water and a set of four electrolytes (NaCl, KCl, $CaCl_2$ and $MgCl_2$) distributed among a set of *p* phases that are characterized by their specific composition with respect to fixed macromolecular polyelectrolytes [JOSEPH, 1971a, b]. The four electrolytes fall into two groups, binary and ternary. The binary group includes NaCl and KCl, characterized as type AB. The other group, comprising $CaCl_2$ and $MgCl_2$, is of the type AB_2. Then, according to equation 11.17:

$$\mu_{AB}' = \mu_{AB}''; \; \mu_{AB_2}' = \mu_{AB_2}''$$

and

$$\mu_{Na}' + \mu_{Cl}' = \mu_{Na}'' + \mu_{Cl}''$$
$$\mu_{Ca}' + 2\mu_{Cl}' = \mu_{Ca}'' + 2\mu_{Cl}''$$

It follows that:

$$\Delta\mu_{Na} = \Delta\mu_{K} = \tfrac{1}{2}\Delta\mu_{Ca} = \tfrac{1}{2}\Delta\mu_{Mg} = -\Delta\mu_{Cl} = \delta \qquad (11.18)$$

where δ is defined as the equivalent change of chemical potential of each kind of ion. In Gibbs' phase rule:

$$f = c + 2 - p \qquad (11.19)$$

where f is the number of degrees of freedom in a system of *c* components and *p* phases, and where the number 2 refers to the conditions of temperature and pressure, when they are variable. This rule is obtained from equation 11.17 when the total number of variable potentials is $p(c - 1)$, related to $c(p - 1)$ conditions of stability or constraint. Then:

$$\begin{aligned} f &= \text{variables} - \text{conditions of constraint} \\ &= p(c - 1) - c(p - 1) \\ &= c - p \end{aligned}$$

Equation 11.19 is obtained when T and P are counted as variables. The entire chain of reasoning may be developed from equation 11.1 or from Gibbs' equation 97. According to the phase rule, the number of degrees of freedom, which is a measure of variability or invariance of a heterogeneous physicochemical system, decreases with the number of phases, *p*. This explains the invariant properties of such species as the large mammals, which are characterized by an extremely large number and variety of cells and tissues, most of which may be heterogeneous at all levels: macroscopic, microscopic and submicroscopic.

In the application to biological systems, the number of degrees of freedom, f, may be defined as the minimum number of independent variables that are necessary to define the equivalent change of chemical potential, δ, when the state of the *milieu intérieur* (blood plasma) is fixed, as in physiological homeostasis. In the standard state of the *milieu intérieur*, it may be regarded, as a first approximation, as a phylogenetic and ontogenetic invariant. Therefore, f refers to the variability of any other given type of cell or tissue in its phylogenetic or ontogenetic development [JOSEPH, 1971a, b].

In general, δ is equal to the ratio of $\Delta\mu_i$ to z_i for any of the five ions, where z is taken as negative for anions. In order to determine f, it is necessary to determine the number of equations necessary to establish $\Delta\mu_i$ for each of the five ions. This is done in the following way. For each ion, there is a fundamental relationship between $\Delta\mu_i$, $(\Delta\mu_i^\circ)'$ and $(c_i)''$:

$$\Delta\mu_i = (\Delta\mu_i^\circ)' + RT \ln \frac{(c_i)''}{(c_i)'} \tag{11.20}$$

where the five values of $(c_i)'$ are fixed by the invariance of the *milieu intérieur.*

The ionization constant, α_i'', is defined by:

$$(\Delta\mu_i^\circ)' = \Delta\mu_i^\circ + RT \ln \frac{\alpha_i''}{\alpha_i'} \tag{11.21}$$

where α' is a constant, characteristic of the *milieu intérieur.*

The dielectric constant, D'', and $\Delta\mu_i^\circ$ are related by five equations of the form:

$$\Delta\mu_i^\circ = f(D'') \tag{11.22}$$

Since α_i'' is also a function of state:

$$(\Delta\mu_i^\circ)' = f'(D'') \tag{11.23}$$

Each ion is described by four parameters ($\Delta\mu_i$, $\Delta\mu_i^\circ$, α_i'' and $(c_i)''$). The ensemble is therefore described by 20 parameters. Equations 11.20–11.23 comprise a total of 19 equations, four of which are given by equation 11.20 and the remainder by the other three sets. An additional condition is required. It is given by the condition of acid-base balance or electrical neutrality:

$$x + \Sigma(c_i)'' z_i = 0 \tag{11.24}$$

There are now 20 equations relating 4 parameters for each of 5 kinds of ions. Two arbitrary parameters, D'' and x, have been introduced as being necessary to establish the physicochemical state. A summary of the necessary conditions is as follows:

1. Four equations to establish δ (equation 11.20).
2. Five equations relating δ to $\Delta\mu_i^\circ$ and α_i''.
3. Five equations relating $\Delta\mu_i^\circ$ to D''.
4. Five equations relating $(\Delta\mu_i^\circ)'$ to D''.
5. One equation relating $(c_i)''$ tf each ion to x.

In order to obtain the necessary 20 equations (4 parameters x 5 ions), it is necessary to introduce two arbitrary parameters: D″ and x. If both of these depend on the chemical morphology, as it has developed phylogenetically and ontogenetically, there is one degree of freedom. Therefore, there is only one normal mode of development for each kind of cell or tissue. Normal growth and aging are therefore channeled into this mode, which represents the standard state of homeostasis. Theoretically, one of the parameters (D″ or x) may remain constant while the other develops in time. This may occur in connective tissue or in red cells [Joseph, 1971a, b]. However, as a general rule, D″ and x change simultaneously in biological systems with the well-ordered and synchronized synthesis of the *n* macromolecular components that characterize each type of structure. In the normal adult state, many cells and tissues tend toward states of invariance. This implies states of both energy and electrolyte balance.

In general, D″ depends on the state of intracellular or extracellular water. The electrical charge, x, is determined by acid-base balance, or by the mean equivalent weight of the macromolecular polyelectrolytes in any phase. Both D″ and x are therefore functions of the chemical morphology of any given type of structure. Configurational free energy and entropy depend largely on D″, which determines all the standard chemical potentials. It also determines dielectric energy and maximal work and tension (chapter IX). Since intracellular rates of metabolism and energy conversion also depend on the intracellular states of aggregation and states of water, all the secondary properties of cells and tissues (X, Y, Z . . .) must depend on the primary functions (A, B, C . . .). The latter set is described by chemical morphology and physicochemical state, which includes the dielectric properties of water and the set of inorganic electrolytes. According to the phase rule, all the primary and secondary properties are highly coordinated and well-ordered univariant functions of chemical morphology.

Chapter XII: Metabolic Energy

The modern concept of energy may be dated roughly from the middle of the 19th century, when it became generally recognized that in all its various forms, the transformations of energy (including mechanical, thermal, electrical and chemical), may be measured and expressed in the same units: calories, joules, foot pounds, and so forth. The general principle of conservation of energy (first law of thermodynamics) is usually dated in the decade 1840–1850, and is credited to JOULE [1843, 1847], VON HELMHOLTZ [1847] and MAYER [1842].

Within the following decades (1850–1900), many physiologists attempted to apply the first law of thermodynamics to animal calorimetry and to the principles of basal metabolism. The relationships of nutrient or metabolic energy to the production of heat and work by the mammalian organism (including man) were also studied in that period. In 1893, RUBNER accurately measured the input and output of energy in dogs. These experiments may be considered to provide the first complete quantitative verification of the first law of thermodynamics in the field of animal calorimetry and metabolism [BENEDICT and CATHCART, 1913]. Working with the respiratory calorimeter, ATWATER *et al.* [1897] extended the observations of RUBNER. By means of complete balance experiments, they proved the validity of the law of conservation of energy, by showing the uniformity of the input and output of nutrient, metabolic and mechanical energy in the human body.

At the time of MAYER's work, according to BENEDICT and CATHCART [1913], the existence of a vital force as a source of physiological force was widely advocated. Among the advocates was LIEBIG [1842], whose views were strongly opposed by MAYER. At all times, there seem to have been many physiologists and other biologists who have explained the origin of physical force in the body as some kind of 'vital force'. This tendency persisted throughout the 19th century, and was opposed by BERNARD [1878], who is thus allied in physiology with MAYER and VON HELMHOLTZ. At the present time, numerous physiologists have given mechanistic rather than thermodynamic interpretations of vital processes. In doing so, they fail to consider the constraints and balances imposed within the animal organism by the operation of the first and second laws of thermodynamics [JOSEPH, 1971a, b]. At the present time, many such 'mechanist' biologists seem to be more or less unconsciously allied with LIEBIG and the other vitalist contemporaries of MAYER and BERNARD. 'It is impossible to suppress the vitalist point of view; it dodges every blow and appears under a new form' [SARTON, 1952].

During this whole period, the common fallacy seems to lie in the assumption that there are internal sources of energy within cells and tissues that are independent of external sources. For example, it has recently been stated on the basis of 'active transport' that 10–30% of the metabolic energy of the body is required to maintain electrolyte balance [GIESE, 1968]. This conclusion is clearly inconsistent with the classical studies on animal calorimetry dating from the period of RUBNER and ATWATER. According to these studies, *all* the nutrient energy is converted to heat, and is quantitatively transferred to the respiratory calorimeter. The body is maintained in a state of constant energy by the quantitative input of nutrient energy [JOSEPH, 1971a, b].

When it is recognized that in the relations between organism and environment, internal energy is never independent of external sources of nutrient energy, it must be evident that all internal vital processes must be constrained by thermodynamic principles, as they are applied to animal metabolism. Thus all transport processes that relate to the repiratory gases and to nutrients and metabolites cannot be independent in the standard state of the organism, but must be well-ordered and coordinated with the transport of heat to the external environment.

The rate of irreversible transport of heat must correspond to the rate of heat formation; this in turn corresponds to the rates of oxygen transport and carbon dioxide production, which are determined by the intracellular oxidative processes. Thus it is quite evident that on the well-established principles of animal calorimetry, *none* of the nutrient energy is required to maintain electrolyte balance in the body. This implies that the net rates of transport of sodium, potassium and chloride are zero in the invariant physicochemical system; the work of transport is also zero [JOSEPH, 1966a].

The concepts of ion 'pumps' and 'carriers' are therefore superfluous (chapter I). Such concepts may be characterized as 'vitalistic' in nature, since they imply the hypotheses of vital forces inherent in cells that do not conform to the laws of thermodynamics. Like many other mechanistic devices, the ideas of pumps and carriers tend to lead to what WHITEHEAD [1925] has called the 'fallacy of misplaced concreteness'. This type of fallacy is avoided in explanations based on the conservation of energy. As has been shown by many of the authorities on the subject, the law of conservation may be based on the general principles of causality; the law is thus completely independent of mechanistic hypotheses, and is never based on ideas of misplaced concreteness.

Causality and Law

'The principle of causality is nothing else but the supposition that all the phenomena of nature are subject to law' [VON HELMHOLTZ, 1847]. Thus it is evident that animal nutrition, respiration and production of heat and work are governed by the laws of energy. Considered in detail, the vital processes that occur within the human body within any 24-hour period are enormously complex and subject to many varied and unpredictable fluctuations. Yet in the broad, general sense, the vital processes are all subject to the laws of thermodynamics, which provide the basis of uniformity which alone enables nutritionists and students of metabolism to arrive at the general laws of bioenergetics. These laws underly all the phenomena of physical biology.

On the subject of causality and law, BERNARD expressed the following views in 1856:

'We seek the laws of phenomena, that is to say, whatsoever is stable, invariable, permanent and eternal in these phenomena. We do not conceive these things except in relation to time, space and number.

'In the immense variety of objects that surround us we thus look for what is invariable, the law. If we see this still varies it is because we have not found the law and we look again for something invariable in the variety. Now we need principles to carry out investigations, ways of thinking. We use our imagination and we represent things abstractly to avoid difficulties. Things are not exactly as in nature, but we are obliged to conceive them thus, and say that we must have ways of proceeding which are more in relation with the nature of things. This signifies nothing because we do not know the nature of things, and these ways of proceeding must be related only to the nature of our own mind.'

BERNARD thus agrees with many others in deferring to a remote future the exact formulation of the laws of nature, as required to establish the principles of causality in biology. The problem has been stated by WHITEHEAD [1925]: 'Obviously, the main recurrences of life are too insistent to escape the notice of the least rational of humans, and even before the dawn of rationality, they have impressed themselves upon the instincts of animals. It is unnecessary to labor the point, that in broad outline certain general states of nature recur, and that our very natures have adapted ourselves to such repetitions.

'But there is a complementary fact which is equally true and equally obvious—nothing ever really recurs in exact detail. No two days are identical, no two winters. What has gone has gone forever. Accordingly, the practical philosophy of mankind has been to accept the details as emanating from the inscrutable womb of things beyond the ken of rationality. Men expected the sun to rise, but the wind bloweth where it listeth.

'Geniuses such as Aristotle, or Archimedes or Roger Bacon must have been endowed with the full scientific mentality, which instinctively holds that all things, great and small are conceivable as exemplifications of general principles which reign throughout the natural order.'

In seeking the laws of phenomena we must find 'whatever is stable, invariable, permanent and eternal'. In recent times, energy and matter have generally been classed among the stable, eternal and invariant. JOULE referred to energy as 'living force'. His paper of 1847 was titled 'Matter, Living Force and Heat', in which he wrote: 'Living force *(vis viva)* is one of the most important qualities with which matter can be endowed, and as such it would be absurd to suppose that it can be destroyed . . . Experiment has shown that wherever living force is *apparently* destroyed, whether by percussion, friction, or any similar means, an exact equivalent of heat is restored. The converse is also true, namely that heat cannot be lessened or absorbed without the production of *living force* or its equivalent attraction through space . . . Heat, living force and attraction through space . . . are mutually convertible. In these conversions nothing is ever lost.'

The matter was expressed by MAYER [1842] as follows: 'Forces are causes, and consequently it is reasonable to apply to them in principle: *causa aequat effectum.*' Also: forces are 'indestructible objects, changing, imponderable'. In the science of ecology, the indestructible forces originate in the solar radiation, are transformed into chemical bond energy and entropy in plants, and are consumed and stored in the bodies of animals. Ultimately they are converted quantitatively by the animal organism to heat and work. Thus at every step the vital processes involve immediate causality in the reversible or irreversible conversion of energy to chemical, mechanical, thermal or other forms. The ultimate cause is the irreversible transfer of radiant energy from the hot sun to the cold earth. This initiates the continuous flow of energy and entropy through all the ecological food chains that exist in the biosphere of the Earth. The biosphere constitutes an enormous open system, which is restricted by conditions of physico-chemical constraint within living organisms. The laws of causality of vital phenomena are based on the energy and entropy principles.

Basal Metabolism

Basal metabolism is essentially a heat function that is a measure of the totality of oxidative reactions in the organism over a given time. It may be measured in calories as the total enthalpy change ΔH per day under resting conditions. It is proportional to the total consumption of fats, carbohydrates and proteins in a 24-hour period. It is also proportional to the daily consumption of oxygen and to the production of carbon dioxide. It also depends on the respiratory quotient (RQ), which is the ratio of CO_2 production to O_2 utilization.

The total enthalpy change, ΔH, is only one aspect of energy metabolism. It measures the heat production under resting conditions; this depends on the totality of all intracellular oxidative processes. In any chemical reaction. ΔH is related to the free energy change ΔG and to the entropy change ΔS. Thus:

$$\Delta G = \Delta H - T\Delta S \tag{12.1}$$

Chemical reactions proceed spontaneously in the direction in which ΔG is negative. Endothermic reactions imply positive values of ΔH. Such reactions can occur spontaneously only when TΔS is positive, and of sufficient magnitude to yield a negative value of ΔG. The intracellular combustion reactions of the body are exothermic. They are also spontaneous and are driven by the negative values of ΔG that are inherent in the thermochemical properties of the nutrients. In the combustion of glucose and other carbohydrates, ΔS is positive. Fats are oxidized with decreases of entropy. The simultaneous oxidation of carbohydrates and fats in definite proportions is required to yield a null value of TΔS. This is a condition of energy balance:

$$\Delta G = \Delta H \tag{12.2}$$

Table XLV. Free energies and heats of combustion at 25 °C (values in kcal/mol, estimated for human physiological conditions)[1]

Reaction	ΔH	TΔS	ΔG
$C + O_2 = CO_2$	–94,025	0.235	–94,260
(1 atm pressure)			
Combustion of glucose			
$Gl + 6O_2 = 6CO_2 + 6H_2O$	–668.5	22.0	–690.5
Combustion of palmitic acid			
$Pa + 23O_2 = 16CO_2 + 16H_2O$	–2,382.1	–45.9	–2,336.2
Formation of lactic acid			
$Gl = 2\,LA$	–17.67	12.21	–29.88

1 JOSEPH [1971a].

Under this condition, the total free energy of the oxidative reactions is converted quantitatively to heat, and transferred irreversibly to the environment. None of the free energy remains in the body to produce additional chemical change or internal work. From the thermochemical constants of table XLV, the conditions of perfect calorimetry or perfect energy balance may be inferred [JOSEPH, 1971a]. In the combustion of 1 mol glucose:

$$C_6H_{12}O_6 + 6\,O_2 = 6\,CO_2 + 6\,H_2O$$
$$\Delta G = -\,690.5 \text{ kcal}$$
$$\Delta H = -\,668.5 \text{ kcal}$$
$$T\Delta S = 22.0 \text{ kcal}$$

In the combustion of palmitic acid:

$$C_{16}H_{32}O_2 + 23\,O_2 = 16\,H_2O = 16\,CO_2$$
$$\Delta H = -\,2{,}382 \text{ kcal}$$
$$\Delta G = -\,2{,}336 \text{ kcal}$$
$$T\Delta S = -\,46 \text{ kcal}$$

Thus to produce perfect energy balance, it is required that TΔS in the oxidation of glucose must be + 46 kcal for each mole of palmitic acid that is oxidized. Thus when b is the number of moles of glucose, and a is the number of moles of palmitic acid, the condition for energy balance is that:

$$\frac{b}{a} = \frac{46.0}{22.0} \tag{12.3}$$

Therefore, under these conditions:

$$2.09\ C_6H_{12}O_6 + 12.54\ O_2 = 12.54\ CO_2 + 12.54\ H_2O$$
$$C_{16}H_{33}O_2 + 23\ O_2 = 16\ CO_2 + 16\ H_2O$$
$$2.09\ C_6H_{12}O_6 + C_{16}H_{33}O_2 + 35.54\ O_2 = 28.54\ CO_2 + 28.54\ H_2O$$

$$\Delta H = -\ 2{,}382 - 1{,}397 = -\ 3{,}779\ \text{kcal}$$
$$\Delta G = -\ 2{,}336 - 1{,}442 = -\ 3{,}779\ \text{kcal}$$
$$T\Delta S = 0$$
$$RQ = \frac{28.54}{35.54} = 0.803$$

The value of the RQ is very similar to the value 0.806 found by SHOCK and YIENGST [1955] for normal men in the age group 40–49 years. Values found for other age groups were 0.811 (50–59 years); 0.827 (60–69 years); 0.802 (70–79 years); 0.808 (80–90 years). The average basal metabolism from 40 to 60 years is about 1,500 kcal/day, measured as ΔH. From 70 to 90 years, the value is of the order of 1,300 kcal or less. When the RQ is 0.827, as in the group from 60 to 69 years, the body is in a state of negative free energy balance: $(-\Delta G)$ exceeds $(-\Delta H)$ by about 7 kcal/day, or a factor of the order of 0.5%. In the other periods, the deviation is of the order of 0.2% or less (3 kcal/day).

Significance of ΔS

Equation 12.1 may be written in the form:

$$\Delta G = \Delta H - T\Delta S_1 - T\Delta S_2 \tag{12.4}$$

where the entropy change ΔS is split into two parts – ΔS_1 and ΔS_2, which signify two kinds of entropy change: chemical and physical. Let us assume that ΔG is $-1{,}501$ kcal/day, and that ΔH is $-1{,}500$ kcal. If there is no change in chemical morphology of the macromolecular components, then $T\Delta S_1$ is zero and $T\Delta S_2$ is 1 kcal. This means that the cells and tissues have expended 1 kcal of free energy in the performance of internal work (for example, in redistribution of ions or water).

In units of mechanical energy:

1 kcal = 4,183 J = 427 kg m
1 kcal = 41 l atm

Thus a value of 1 kcal implies sufficient energy to lift a man of 100 kg to a height of 4.27 m. It also implies sufficient energy to compress about 21 l of air at 0 °C and 1 atm pressure to a pressure of 10 atm. Expressed as internal work, a figure of 1 kcal for TΔS appears to be either inconceivable or highly improbable. Nevertheless, figures as high as 10–30% of the total respiratory metabolism have appeared in the literature as required to maintain electrolyte balance by means of 'active transport'. This would imply values of the order of 150–450 kcal/day for TΔS. Then when ΔH is – 1,500 kcal and TΔS is 450 kcal, the value of ΔG would be – 1,950 kcal/day, implying a negative free energy balance of 450 kcal.

In the combustion of 1 mol of glucose, TΔS is 22.0 kcal, and ΔH is – 668.5 kcal. To yield 1,500 kcal, the combustion of about 2.25 mol of glucose is required. This yields a value of 49.5 kcal for TΔS, which is much less than the value of 150–450 kcal required for the operation of a 'sodium pump'.

The estimated figure for the internal work of maintaining electrolyte balance is clearly incompatible with the known values of ΔH (– 1,500 kcal) and of the RQ (0.80–0.83). The true figure of TΔS calculated from basal metabolism and basal free energy is of the order of zero, and at most amonts to 5 kcal/day, rather than 150–450 kcal, as calculated from 'active transport'. The assumptions underlying the theory of active transport thus require sources of internal energy (or 'vital force') that are of magnitudes beyond those available from nutrient sources. On the basis of internal work (or $T\Delta S_2$), the theory is clearly incompatible with the principles of basal metabolism and animal calorimetry.

Whereas a magnitude of 1 kcal for internal work is inconceivable, the same magnitude is not inconceivable for morphological change. At a temperature of 37 °C, the value of ΔS_1 would be 3.2 entropy units when $T\Delta S_1$ is 1.0 kcal. The entropy of glucose is 0.28 units/g; for fat the value is 0.44 units/g. An increase of 10 g of fat in the body, accompanied by a decrease of 10 g of carbohydrate would represent an increase of 10 (0.44–0.28), or 1.6 entropy units. This connotes a value of 0.5 kcal for $T\Delta S_1$, which is a moderate degree of negative free energy balance (about 0.03% of ΔH).

Internal and External Work[1]

In the adult period of mammals, including man, the body is an invariant system on a day-to-day basis. Perfect invariance implies that on such a basis, $\Sigma\Delta S$ is zero for nutrient metabolism and that $\Sigma\Delta G = \Sigma\Delta H$.

Over short-term periods, when the body performs vigorous muscular work, it is in a state of negative free energy balance. In a state of basal metabolism, a typical value of ΔH is of the order of $-1{,}500$ kcal/day, or -62.5 kcal/h. During the performance of vigorous muscular work, the figure may increase to a higher order of magnitude. For example, the value of ΔH may become $-4{,}000$ kcal/day, or -166.7 kcal/h. A state of perfect energy and electrolyte balance on a day-to-day basis then implies the value of $-4{,}000$ kcal/day, an increase of 2,500 kcal in the output of free energy and of heat. This meets the condition of perfect energy balance, which requires that $\Sigma\Delta S = 0$. It is also required, however, that sufficient nutrients and oxygen are taken in to supply 4,000 kcal of enthalpy and of free energy.

Let it be assumed that the excess output of heat occurs over a period of 10 h; this implies a metabolic rate of (250 + 62.5) or 312.5 kcal/h. This is not excessive for a healthy adult man. During this rather prolonged period of physical activity, the body would be in a state of negative free energy balance. In muscular activity, free energy and 'negative entropy' are supplied by the anaerobic breakdown of glucose to lactic acid, and by the oxidation of the products of glycolysis. This is accompanied by an increase of the respiratory quotient (RQ). If during this period, the RQ increases to a mean value of 0.82, the ratio of b to a, which represents the ratio of moles of oxidized glucose to moles of oxidized palmitic acid, attains a value of about 2.65 [Joseph, 1971a]. This implies a value of $(-\Delta G)$ that exceeds $(-\Delta H)$ by a factor of about 0.3%. Thus during the 10-hour period in which ΔH is -312.5 kcal/h, the calculated value of ΔG would be about -313.5 kcal. In the 10-hour period of vigorous work:

1 The following calculations are entirely hypothetical, and are intended only to show the nature of intracellular metabolism, morphological changes and electrolyte balance in active and resting muscle. Energy production and utilization depend on changes of state of intramuscular contractile proteins, water and electrolytes. Changes of state are largely reversible, and would fluctuate widely from time to time in the hypothetical 10-hour work period. For simplicity, it is assumed that metabolic rates and RQ are constant during the period of muscular work. Actually, this would never be true except as a hypothetical first approximation, in which external work is performed at a perfectly uniform rate.

$\Delta H = -\ 3{,}125$ kcal
$\Delta G = -\ 3{,}135$ kcal

In the ensuing 14-hour period during which ΔH is $-\ 62.5$ kcal/h:

$\Delta H = -\ 875$ kcal

$\Delta G = -\ 865$ kcal

$T\Delta S = -\ 10$ kcal

In the 10-hour period:

$\Delta H = -\ 2{,}382\ a\ -\ 668.5\ b\ =\ -\ 3{,}125$ kcal

$\Delta G = -\ 2{,}336\ a\ -\ 690.5\ b\ =\ -\ 3{,}135$ kcal

$T\Delta S = 10$ kcal $\quad \Delta S = a\,\Delta S_a + b\,\Delta S_b$

$T\Delta S = -\ 46.0\ a + 22.0\ b$

where $-\ 46.0$ kcal is the value of $T\Delta S_a$ (palmitic acid), and where $+\ 22.0$ is the value of $T\Delta S_b$ (glucose). From the value of 0.82 for RQ, $b/a = 2.65$. Hence:

$-\ 46.0\ a + 22.0 \times 2.65\ a = 10$
$a = 0.81;\ b = 2.15$

This implies that during the 10-hour period of muscular activity, the body is in a state of negative free energy balance amounting to 1 kcal/h. This represents about 0.3% of the value of $(-\Delta H)$. It implies a loss of configurational free energy corresponding to $-\ \Delta G_c$ or to $T\Delta S_c$. In this period, the value of $c_{Na}''\Delta\mu_{Na}^{\circ}$ or RT $[\Sigma c_{i'} - \Sigma(c_i)'']$ would decrease in each kind of active muscle, and the corresponding values of D″ would increase. According to the calculations of chapter IX, D″ increases during muscular contraction; the water structure is broken, the metabolic rate increases, and there is an increase of CO_2 production, O_2 consumption and of the RQ. The concomitant anaerobic breakdown of glucose to lactic acid is activated by the change of state of the intrafibrillar water; this increases the supply of 'negative entropy' that is necessary to repolarize the fibers during the recovery process after each contraction. This process maintains the standard action and resting potentials at the expense of glycogenolysis and anaerobic glycolysis. However, 'negative entropy' is lost from the body during the period in which TΔS increases at the rate of 1 kcal/h. During this period, the RQ is high, and the body is in a state of imbalance, in which (−ΔG) exceeds (−ΔH) by 10 kcal in a 10-hour period.

To maintain perfect energy and electrolyte balance over a period of 24 h, it is necessary that the value of TΔS become −10 kcal over the remaining 14 h. During this resting period:

$$\Delta H = -\ 875 \text{ kcal}$$

$$\Delta G = -\ 865 \text{ kcal}$$

$$T\Delta S = 46.0\ a + 22.0\ b = -\ 10 \text{ kcal}$$

$$T\Delta S = 46.0\ a \pm \mathit{22.0}\, b = -\ 10 \text{ kcal}$$

$$a = 0.370;\ \ b = 0.192;\ \ b/a = 0.51$$

$$RQ = \frac{16\ a + 6\ b}{23\ a + 6\ b} = 0.73$$

In order to compensate for the 10-hour period of muscular activity, in which (−ΔH) averages 312.5 kcal/h, it is necessary for the oxidative metabolism to shift from a high rate of carbohydrate utilization (RQ = 0.82) to a 14-hour period in which fat utilization is high (RQ = 0.73). Then over the 24-hour period, TΔS is zero, and − ΔH is equal to − ΔG (4,000 kcal). During the period of muscular activity and high utilization of carbohydrate, the metabolic rate is high (312.5 kcal/h). Corresponding to the state of high entropy, intracellular water tends toward a state of high dielectric constant. This stimulates anaerobic glycolytic metabolism, with the production of 'negative entropy'. This maintains the well-ordered state of low entropy in the system of contractile fibers, water and inorganic electrolytes. Nevertheless, the loss of free energy in the period of energy imbalance implies changes in chemical morphology and dielectric energy.

In the ensuing rest period, the system shifts to a state of high fat metabolism, in which the RQ is low, and during which muscle is in a state of positive free energy balance (TΔS = −10 kcal in the 14-hour period.) In the recovery process, dielectric energy and normal electrolyte balance are restored, and the changes in chemical morphology are reversed. This is a period in which fat metabolism is high, and in which the dielectric constant is low. In this well-ordered state of intramuscular water, transport of fat into the sarcoplasm would correspond to a high rate of oxidation;

the permeability of the sarcolemma or cell wall accomodates itself to the high rate of utilization of the fat [Fritz, 1961]. Conversely, when the intracellular dielectric constant is high, intracellular water behaves as a polar rather than as a non-polar solvent. In this state, transport of glucose and other water-soluble metabolites such as amino acids would be facilitated, and transport of fats and other lipids would be suppressed.

According to this view of metabolism, muscle responds with one degree of freedom to changes of contraction and tension. Metabolic rates and respiratory quotient are functions of the state of order or disorder in the aqueous dielectric. This determines the nature of the internal metabolic processes and the rates of transport of nutrients and metabolites. Permeabilities and other transport properties are secondary properties (chapter I). They are determined by the dielectric properties of each part of the cell. These properties depend on the physiological and metabolic properties of the cell as a whole. During stimulation, dielectric energy is redistributed throughout all parts of the system. No part of the system can react independently of all the other parts.

In the above example of muscular metabolism, the value of $T\Delta S$ taken over a 24-hour period is zero when the final state of the body is the same as the initial state. In the hypothetical period of muscular work, $T\Delta S$ is 10 kcal, corresponding to a small negative free energy balance. In the 14-hour recovery period, the value of $T\Delta S$ is -10 kcal, indicating that the body has returned to its initial standard resting state. Thus over any time interval during which the body returns to its initial standard state, $T\Delta S$ is zero. This requires the general condition of invariance that $\Delta G = \Delta H$. It is implied then that the total free energy of all oxidative reactions has been converted to heat, and that none remains to produce changes of chemical morphology, physicochemical state, or changes of temperature. This principle will be shown to have important implications in cardiovascular physiology [Joseph, 1971a].

Rates of Glycolysis

When skeletal muscle is stimulated in the absence of oxygen, it is capable of performing prolonged work under various conditions. It is possible, by purely anaerobic processes, for it to undergo repeated isotonic contractions which yield external work; it is also possible under faradic stimulation to develop considerable tension (3,000 g/cm^2) under isometric conditions. In either case, anaerobic contraction involves the production of lactic acid and heat. The ultimate source of the formation of heat is the breakdown of glycogen to lactic acid. In isometric contraction, the free energy change, ΔG, of the process is converted to heat rather than to work. The reaction proceeds through the formation of the ester, glucose 1-phosphate. The breakdown of this ester proceeds through the successive formation of a series of phosphate esters, which include: fructose 6-phosphate; fructofuranose 1-phosphate; fructofuranose 1,6-phosphate; phosphoglyceric aldehyde; phosphoglyceric acid and intermediate steps, leading to: phosphopyruvic acid, pyruvic acid and lactic acid.

The series of reactions in muscle and yeast are very similar up to the point of the formation of pyruvic acid. Unlike yeast, muscle does not contain carboxylase. In yeast, alcoholic fermentation proceeds through steps involving carboxylase, but in muscle, anaerobic glycolysis terminates with the formation of lactic acid.

The anaerobic process of muscular work is described by the following processes: (1) ATP does not change greatly; (2) creatine phosphate breaks down; (3) inorganic phosphate increases; (4) free creatine appears; (5) lactic acid is produced.

After the end of the active period, a process of anaerobic recovery ensues, in which phosphagen (creatine phosphate) is resynthesized (reversal of step 2) at the expense of inorganic phosphate and free creatine (steps 3 and 4).

Phosphorylation of many of the intermediate steps from glycogen to lactic acid occurs in the reversible breakdown of ATP to ADP. However, the level of ATP is maintained by resynthesis at the expense of phosphocreatine. This is known as the Lohmann reaction:

$$CP + ADP = ATP + C$$

No inorganic phosphate is liberated, and the energy change is small. The bulk of inorganic phosphate formed in anaerobic contraction occurs in the breakdown of phosphagen. The work of muscular contraction is produced at the expense of an increase of configurational entropy, resulting from a change of state of the intrafibrillar water (chapter X).

The rate of anaerobic glycolysis, which measures the formation of lactic acid, heat and work therefore corresponds to the rate of formation of creatine and inorganic phosphate, when ATP and ADP remain in steady states. This is possible until the onset of muscular fatigue.

Effect of Dielectric Constant

When intracellular water passes from a state of low dielectric constant, D″, to a state of disorder and high dielectric constant, the effect is to stimulate the rate of anaerobic glycolysis. The initial step is the formation of α glucose 1-phosphate from glucose or glycogen:

$$\text{Glucose} + \text{ATP} = \text{ADP} + \text{glucose 1-phosphate} \quad (1)$$

This is the rate determining process for the breakdown of phosphagen:

$$\text{CP} + \text{ADP} = \text{ATP} + \text{C} \quad (2)$$

A summation over the two reactions yields:

$$\text{Glucose} + \text{CP} = \text{glucose 1-phosphate} + \text{C} \quad (3)$$

According to the theory of reaction rates, the kinetics of reaction 3 at constant dielectric conditions depend on the activation energy, E_a. This is related to the free energy of activation, ΔG^*, described by the relation:

$$\Delta G^* = -\text{RT} \ln \text{K}^* \quad (12.6)$$

where

$$\text{K}^* = \frac{(\text{GPC})^*}{(\text{G})(\text{CP})} \quad (12.7)$$

In the numerator $(\text{GPC})^*$ denotes the quantity per unit volume of the ‘activated complex’ formed by the combination of glucose and the univalent anion, creatine phosphate. The activated complex $(\text{GPC})^*$ is an unstable intermediary between the reactants and glucose 1-phosphate. The rate of its breakdown is a measure of the rate of formation of free creatine – hence a measure of the rate of anaerobic glycolysis. In the theory of reaction rates, the rate of reaction 3 may be denoted as k_r. It is proportional to K^*:

$$k_r = \left(\frac{kT}{h}\right) K^* \quad (12.8)$$

and

$$k_r = \left(\frac{kT}{h}\right) \exp\left(\frac{-\Delta G^*}{RT}\right) \quad (12.9)$$

where k is the Boltzmann constant, and h is Planck’s quantum of action.

The free energy of activation is related to the entropy of activation, S^* and the heat of activation, H^*. Hence:

$$k_r = \left(\frac{kT}{h}\right) \exp\left(\frac{-\Delta H^*}{RT}\right) \exp\left(\frac{\Delta S^*}{R}\right) \tag{12.10}$$

Equations 12.7–12.10 represent the kinetics of reaction 3 in a medium of dielectric constant D″. All the conditions of reaction kinetics would change when the intracellular or intramuscular water undergoes a change of state, as in muscular contraction. The initial effect of changing water from a standard structured 'ice-like' form to a liquid state of disorder would be to solubilize substances such as glycogen, glucose and creatine phosphate. This would increase the concentration of the activated complex (GPC)*, to which the reaction rate is theoretically proportional. In the state of high dielectric constant, this concentration may be represented as $(\overline{GPC})^*$. Then:

$$\frac{\bar{k}_r}{k_r} = \frac{(\overline{GPC})^*}{(GPC)^*} = P$$

Thus, in these simple terms, the rate constant $\bar{k}_r$ is increased over k_r by a factor of ρ in any change of state of water that results in increases of solubility of glycogen, glucose and phosphagen. The metabolic rate is thus easily understood to be a function of the intracellular state of water, which determines the solubilities and concentrations of all solutes and activated complexes. These fluctuate with the states of water, returning to the standard basal states only with a return of the water, electrolytes and macromolecular polyelectrolytes to the physiological standard state. Over any period during which the system returns to the standard state, the respiratory rate and respiratory quotient depend on the internal state of the dielectric in a one-to-one relationship. Over such a period in which the initial and final states of the dielectric are the same, the thermodynamic functions such as free energy, entropy and chemical potentials are expressed as line integrals with values of zero. This is the thermodynamic condition for a conservative system which, aside from secular fluctuations, is invariant on a day-to-day basis. Respiratory rates and quotients also fluctuate with heat production and transport of nutrients and metabolites, but the fluctuations occur under conditions of constraint that maintain day-to-day invariance in the system as a whole.

In the foregoing conceptual scheme of glycolysis and isotonic or isometric contraction, the function of ATP is to react in the formation of glucose 1-phosphate and of several other phosphate esters. The splitting of the high energy phosphate bond does not supply the energy for external work or for the production of a state of isometric tension. These are related to the changes of configurational free energy and entropy of water, electrolytes and contractile proteins. In particular, the change of configurational free energy is proportional to:

$c_{Na}'' \Delta\mu_{Na}^{\circ}$ and to RT $[\Sigma_{ci}' - \Sigma(c_i)'']$ (chapter IX).

During a series of anaerobic muscular contractions, the conversion of ATP to ADP is reversible, and each substance remains in a steady state. The irreversible processes are the breakdown of glycogen and glucose to lactic acid and the breakdown of phosphagen to inorganic phosphate and free creatine. These occur during the periods when the dielectric constant is high. Reversal to a state of order and a state of basal metabolism requires a supply of negative entropy; this implies aerobic respiration, a state of low dielectric constant, and mitochondrial oxidation of fats and other nutrients. Since ATP remains in a steady state whenever muscle is in its standard state, it does not supply energy for work or tension.

Intracellular respiratory metabolism under either aerobic or anaerobic conditions thus depends on the state of water as a dispersion medium. In a state of high dielectric constant, the well-ordered water structure is broken. Then water behaves as a polar solvent, in which glucose and glycogen become soluble. In the well-ordered state of low dielectric constant, it behaves as a non-polar solvent for fats and other lipids. Transport of fats, carbohydrates and other nutrients from extracellular to intracellular phases depends on the state of water in all parts of the system. Coordination of transport with respiratory metabolism implies coordinated changes of state throughout the system. This signifies that transport is controlled by the metabolic processes as a whole. Transport is thus a thermodynamic rather than a mechanical process. To explain it by mechanistic models is to submit to the fallacy of misplaced concreteness.

Transfer of Dielectric Energy

Through the system of interfibrillar, intrafibrillar, intermuscular and membranous structures, muscles exert tensile forces on other muscles and tissues. In this way, mechanical energy, tension and dielectric energy are transferred from one structure to another; configurational energy and entropy are at all times continually redistributed throughout entire organs and tissues.

When one set of muscles, A, is stretched, another set, B, may be contracted. In the contracted state, either isotonic or isometric, the dielectric constant of B is high; the rate of anaerobic glycolysis and the RQ are also high, corresponding to a period of negative free energy balance. At the same time, in a period of negative free energy balance intramuscular water is in a state of high dielectric constant, and the muscle is in a state of tension. At any time, the situation may be reversed, and set A may go into a period of contraction while set B is stretched. In this way, tension, length, metabolic rates and dielectric energy may be continually redistributed throughout a system of muscles, myofibrils and connective tissues.

In any myofibril, dielectric energy is measured by the product c_{Na}'' $\Delta\mu_{Na}^{\circ}$. It thus depends on the dielectric constant D″. This changes with the intrafibrillar state of water. When D″ is high, the myoplasm tends to behave as a polar solvent. It transports and metabolizes water-soluble substances such as glucose and amino acids. It lowers the standard chemical potentials and increases the solubilities of sodium and other inorganic ions. In this state, electrical polarization is low, and electrical potentials approach the value of E_{Na}, the 'spike potential' of the fiber.

On the other hand, when $c_{Na}'' \Delta\mu_{Na}^{\circ}$ is in the standard resting state, the myoplasm tends to behave as a non-polar solvent, which metabolizes and transports fats and other lipids. Mitochondrial oxidation is stimulated, anaerobic glycolysis is suppressed or reversed, and the RQ is lowered. In the state of low dielectric constant, the standard resting condition of the fiber is restored; a state of low dielectric constant of the intramuscular dispersion medium corresponds to a period of high electrical polarity between intracellular and extracellular phases. This is a measure of the irritability, tone and responsiveness of the cellular phases.

Thus there is a high correspondence among the following properties of skeletal muscle and other reactive tissues; mechanical tension, dielectric energy, metabolic rates and type of metabolism, and production of heat and mechanical work. All of these depend on the mechanical state and chemical morphology of the intracellular and extracellular phases, which may vary continually in both resting and active periods. Thus dielectric energy may be continually redistributed throughout an entire system under conditions in which the total energy remains constant. In this way one set of muscle fibers may influence all the properties of another set. Fibers therefore do not metabolize independently of each other. The metabolism of entire sets of fibers is controlled by the mechanical state of the entire ensemble. Growth, metabolism and electrical properties depend on the physicochemical state and activity of the organism as a whole. All the properties of any part of the system change with the polarity of the liquid parts of the dielectric, which determine its solvent properties and the type of metabolism.

Work of the Heart

According to STARLING [1912], the total work of the adult human heart is of the order of 10,000 kg m or 100,000 J/day. Beating at the rate of 70 strokes/min, the total number of contractions per day is of the order of 100,000; the work per contraction is then of the order of 1 J or 0.24 cal/beat. However, within the period of 1 heartbeat, all this energy is converted to heat and transported to the environment. This is a condition for perfect energy balance within the body. It requires that: $\Delta G = \Delta H$, and that $T\Delta S = 0$.

These are the general conditions for thermal, physicochemical, and mechanical invariance. During systolic contraction, pressure and kinetic energy are increased in the cardiovascular system. In the interval of one cardiac cycle; however, the values return to the diastolic levels before the process is repeated. The condition for this is that the diastolic level of cardiovascular energy must return to its characteristic level. The function of the heart is to maintain constant conditions in the circulatory system, rather than to perform irreversible work. This requires that it maintain the pressure gradients necessary to overcome peripheral resistance in the capillaries and connective tissue ground substance. Somatic cells and tissues require a continuous flow of tissue fluid through the ground substance. This depends on the maintenance of a steady pressure gradient in the system of capillaries that transport the arterial blood to the small veins.

This gradient is the force that causes flow of tissue fluid through the water-rich phase of connective tissue. Hydrostatic resistance in the ground substance depends on the state of its organization with respect to the system of submicroscopic vacuoles which contain the bulk of the tissue fluid. This system depends on the aging of ground substance, and on the physicochemical state of collagen, glycoproteins and mucopolysaccharides [BONDAREFF, 1957; GERSH and CATCHPOLE, 1960; JOSEPH, 1971b]. Therefore, the cardiovascular hypertension of senescence may depend, in large part, on the aging of connective tissue. This would tend to increase peripheral resistance and pressure gradients in the capillary circulation. Secondary effects would result in increases of systolic and diastolic blood pressure and in increased work of the heart.

Functions of Water

In the poetic ages of history, the more articulate voices of mankind have been possessed of the direct perception of nature. The community of living things, which in the modern scientific scheme, is studied under the subject of ecology or physical biology, was then the province of the great poets: Virgil, Lucretius, Chaucer and Shakespeare. Thus in the prologue to the 'Canterbury Tales':

> 'When that April with his showrs soote
> The drought of March had perced to the roote
> And bathed every vein in swich licour
> Of which vertu engendered is the flowr.'

At all times, the poet of Nature has been fully aware of the central function of the rains in the life of the Universe. In this he is at one with all those who commune intimately with nature: the agriculturists, the artists and the naturalists.

Laboratory scientists, on the other hand, have often shown an unfortunate tendency to isolate themselves from direct contact with living nature, and to lose the direct perception of the poet, artist and musician.

Thus, the elaborate schemes of enzymology and intermediary metabolism seem to have taken little or no account of the functions of intracellular, extracellular and environmental water, which vary with physicochemical state. Because of their isolation in the laboratory or classroom, many academic biologists seem to have shielded themselves from the April rains and from:

> '...Zephyrus, eek, with his sweete breeth
> Inspired hath in every holt and heeth
> The tender croppes, ...

The same laboratory biologist, who has so enriched our knowledge of intermediary metabolites (which now probably are numbered in the four-figured class of integers), has barely acknowledged that the functions of all these substances and of the system as a whole depend on the 'showrs' which bathe all the cells and tissues of plants and animals.

BERNARD, who flourished in the 19th century, certainly was not a victim of this typical myopia. It was he who emphasized the central function of the *milieu intérieur* in bathing all the cells and tissues in a constant culture medium, which isolates all of them from the rigors of exposure to the external physical world. In this he was followed by LOEB, MACALLUM, HENDERSON, CANNON and many other biologists and ecologists who were conscious of the life of organisms in a state of nature, in addition to being objects of laboratory experimentation. In the state of nature, the life of every cell and tissue depends on nutritional sources of metabolic energy. None of the cellular energy can be conceived to be of purely internal origin. All of it originates in the cells and tissues of other organisms. The life of the whole depends on the constant supply of energy and entropy that originates in the Sun. In the economy of the whole, the rains and sunlight play predominant roles. Directly, they activate the life of the entire vegetable kingdom. Directly or indirectly, they sustain and enrich the lives of all other species.

Chapter XIII: Sets and Nomograms

In the conceptual scheme of physical biology that has been presented in the preceding chapters, chemical morphology has been treated as the primary function on which all secondary properties depend. Chemical morphology of any biological structure is described by means of certain sets and subsets of chemical substances. Any biological structure may be regarded as a well-ordered aggregate M which consists of a number of elements, m. For example, in mammalian skeletal muscle, M may include a well-defined and well-ordered group of proteins of various kinds, denoted as P. The elements of P may be denoted as $p_1, p_2, p_3 \ldots$ and would include actin, myosin and other muscle proteins.

The aggregate M also includes other sets of components: water and inorganic ions, treated as a set; carbohydrates such as glycogen, glucose and a large number of intermediary substances involved in anaerobic glycolysis may also be conceptualized as sets. Other sets or classes of metabolites include organic phosphates: phosphagen, ATP, ADP and those formed from hexoses, pentoses and trioses. Amino acids, peptides and proteins form sets which interlock at certain points with other sets of metabolites. Thus lactic acid may be formed by deamination of alanine; such exchanges of nitrogen or amino groups may be reversible. Threonine, aspartic and glutamic acids may also be deaminated, and may have glycogenetic and respiratory functions.

In biological systems, physicochemical or physiological behavior can only be conceptualized as the ordered responses of entire sets of chemical substances which pass from one well-ordered state (A, B, C . . .) to another well-ordered but modified state (A′, B′, C′ . . .). In the normally well-ordered response, the system may show only one degree of freedom. According to the terminology of the phase rule, it is univariant. This implies that in a change of state of one substance A to A′ there must be coordinated changes of B, C . . . to include all the other members of the set. Thus if:

$$M = (m_1, m_2 \ldots)$$

then

$M' = (m_1, m_2 \ldots)$

In the description of cells and tissues by means of sets, it is often the case that a given substance, S, may be a member of two or more different sets. Thus lactic and pyruvic acids, in addition to being members of the set of intermediary metabolites in aerobic or anaerobic metabolism, also belong to the set of intracellular acids and bases; therefore they have functions in intracellular and extracellular acid-base equilibrium. This is also true of practically all intracellular and extracellular proteins, peptides and amino acids; they can all combine with or dissociate hydrogen ions, and therefore enter into acid-base balance.

The respiratory gases, oxygen and carbon dioxide, form a simple two-membered set that enters all phases of cells and tissues. Each gas enters into the behavior of other sets of substances, in particular by affecting acid-base balance and redox potentials. Thus each gas affects the physicochemical behavior of the other [HENDERSON, 1928]. Carbon dioxide tension affects the oxygen-carrying properties of blood; oxygen tension changes the carbon dioxide dissociation curve. Oxyhemoglobin is a more strongly acidic protein than is reduced hemoglobin. Oxygen affects acid-base balance by acting on this property. Carbon dioxide forms carbonic acid and bicarbonate ions in blood; the equilibrium depends on the content of intracellular and plasma proteins, on the electrolyte balance, and on the oxygen tension. The complex interplay of physicochemical forces acting in the heterogeneous system: plasma, erythrocytes, lung and connective tissue can only be understood as the correlated univariant behavior of entire sets of properties. When this behavior is formulated by means of thermodynamic or phase-rule principles, it is then possible to present it graphically by means of contour diagrams or by Cartesian and d'Ocagne nomograms. HENDERSON [1928] has given simple directions for the construction of such diagrams, by means of which it is possible to show the behavior of very complex physiological systems by various kinds of two-dimensional graphical diagrams.

Characteristics of Living Systems

Any living animal falls naturally into a well-defined phylum, order, class, genus and species. Thus during the entire life span of higher organisms, the morphology and behavior may be described by certain invariant properties which distinguish it from animals of other species which show other types of invariant properties. However, all species of multicellular animals, or metazoa, may be described by a certain number of essential properties which distinguish them from non-living forms.

First: higher organisms with sufficiently long life spans show in the adult period the property of day-to-day invariance.

Second: the organisms under specified conditions convert nutrient energy at characteristic standard rates.

Third: the organisms are well-ordered, structured and heterogeneous. It may be shown that this is the necessary condition for day-to-day invariant morphology and reaction rates.

Thus it can be said that no unstructured or homogeneous organism can be both reactive and invariant. Reactions in homogeneous systems must necessarily lead to physicochemical changes; the rates depend on the state at any given instant of time.

The unique properties of living systems illustrated by invariant structure, composition and metabolic rates are diagrammed in figure 2. The three circles, I, II and III, form a region IV, which has properties common to

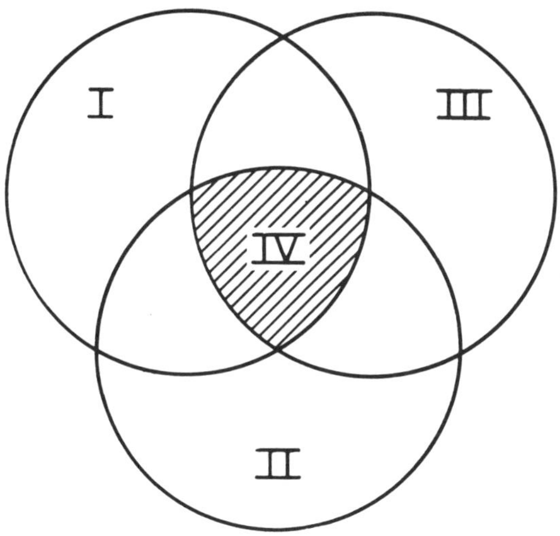

Fig. 2. Physicochemical conditions for invariance and reactivity of biological systems. I: set of all invariant systems; II: set of all heterogeneous systems; III: set of all reactive systems; IV: set of all systems that are invariant, heterogeneous and reactive. All living cells and tissues are in set IV.

each of the three circles. This area, to which living organisms are confined, includes the class or set of all systems that are (I) invariant, (II) heterogeneous, and (III) reactive.

Another area of the diagram common to I and II, includes the set of all systems that are invariant, heterogeneous and *non-reactive.* It would then include anatomical or histological sections that represent the morphology of a non-living biological system, which undergoes no chemical change or reactions.

A third area of the diagram, common to II and III, represents the class of systems that are heteregeneous, reactive, but not invariant. Biochemical preparations that are heterogeneous and structured, but which are not constrained by thermodynamic conditions of invariance, would fall into this class.

Finally, the area of the diagram that is common to I and III, represents a null or non-existing class of systems that are homogeneous, invariant and reactive. In this null set the conditions of homogeneity and reactivity preclude the possibility of invariance.

Sets and Aggregates

According to the foregoing, the description of biological systems requires the application of the logic of classes, sets or aggregates. The classification of any biological system as a physicochemical system is the first step in finding the laws of its structure and behavior. Accordingly, it will be necessary to present some of the essential principles of set theory as a background for the theoretical description of systems that are encountered in physical biology.

The concept of 'sets' is taken as primitive and as synonymous with those of 'classes' or 'aggregates'. The terms 'entity', 'element', or 'object' are also primitive; no attempt is made to define these terms in relation to simpler concepts. According to CANTOR [1895, 1897], 'By an aggregate *(Menge)* we are to understand any collection into a whole *(Zusammenfassung zu einem Ganzen)* M of definite and separate objects of our intuition or our thought. These objects are called the 'elements' of M.

'In signs we express this thus:

$$M = (m)$$

'We denote the uniting of many aggregates M, N, P . . . which have no common elements into a single aggregate by:

(M, N, P . . .)'

The development of mathematics from these primitive logical principles was largely the work of CANTOR, FREGE, PEANO, WHITEHEAD and RUSSELL, and many others [VAN HEIJEVOORT, 1967]. Among the most important results were the establishment of the theory of transfinite numbers and the theory of cardinal and ordinal numbers. In biology, the problem is not that of pure analysis in terms of mathematical and logical concepts. In pure analysis, the problem is that of establishing one-to-one correspondences between abstract mathematical points in a space-time continuum. The biological problem is that of establishing one-to-one biunivocal correspondences between a limited number of well-defined (but usually discontinuous) entities, the magnitudes of which are measurable in physical or chemical units. These differ from the 'entities', 'elements' or 'objects' of pure mathematics or logic. Nevertheless, it is possible to establish biunivocal correspondence among the measurable quantities that constitute the real and characteristic sets of biological systems. This involves geometrical transformations of the measurable magnitudes into alignment charts or 'mappings' of the ordered sets. Conveniently, these may take the form of Cartesian or d'Ocagne nomograms [HENDERSON, 1928].

In preceding chapters, the problem of coordinating the various mathematical sets of a biological system was resolved by means of the thermodynamics of heterogeneous systems – in particular by the application of the phase rule. Here, the measurable physicochemical quantities are expressed by the sets of thermodynamic functions. Thus for each member of the set of inorganic ions (sodium, potassium, calcium, magnesium and chloride), the following quantities are treated as real, measurable or calculable:

1. The change of chemical potential, $\Delta\mu_i$.
2. The change of standard chemical potential, $\Delta\mu_i^\circ$.
3. The total concentration, $(c_i)''$.
4. The ionization constant, α_i''.

Thus each of the five ions is characterized by four properties – a total of 20 variables. Given a standard state of the *milieu intérieur,* the system has only one degree of freedom. This is determined by the chemical morphology of the given cell or tissue, taken as the independent variable.

We are now involved in an important question relating to mathematical terminology. What is the argument or *independent variable?* FREGE [1952] dealt with this question in an essay 'What is a function?'. He showed that the logical and methodological questions involved in the definition of the concepts of number, function and variable are extremely difficult. A variable is a quantity that changes – all change occurs in time. Applying this thought to biological systems, any physically or chemically measurable variable is a function of time. Therefore it depends on growth, development and aging. What is it that develops in time? It is the organism as a whole, which can be described as the set S of all its subsets. It appears then that one element of S, for example, intracellular potassium concentration $(c_K)''$, in a specified region depends ultimately on S, which is the set of all sets (M, N, O, P . . .), that characterizes the organism as a whole. If P is the set of all muscle proteins, $p_1, p_2, p_3 \ldots$ then in particular:

$$(c_k)'' = f(p_1, p_2, p_3 \ldots)$$

But

$$S = (M, N, O, P \ldots)$$

Therefore:

$$S = ((c_k)'' \ldots)$$

Viewed in this way, the analysis of chemical morphology can always be shown to depend on S, the state of the organism as a whole. Thus each of the subsets M, N, O, P . . . stands in biunivocal correspondence with each other. There is a one-to-one correspondence between each of the ions, Na, K, Ca, Mg and Cl that describe the standard chemical morphology of each kind of cell and tissue. The entire system is well-ordered. No individual kind of structure develops independently of the entire ensemble.

Alignment Charts

An equation that relates three variables: (x, y, z; x, y, u; x, u, z, and so forth) may always be represented graphically in the form of two-dimensional *contour line chart.* A set of equations involving such families of variables can be represented graphically either by Cartesian or by d'Ocagne nomograms [HENDERSON, 1928]. For example, if:

$$z = x + y$$
$$u = x - y$$

each equation corresponds to a contour line chart, which may be represented in the form of a Cartesian nomogram. In the study of blood as a physico-chemical system, seven elementary functions may be expressed in the following form:

$$\text{Total } CO_2 = f_1\ (pCO_2, pO_2)$$

$$\text{Total } O_2 = f_2(pCO_2, pO_2)$$

$$[H^+]s = f_3(pCO_2, pO_2)$$

$$r = f_4(pCO_2, pO_2)$$

$$v = f_5(pCO_2, pO_2)$$

where pCO_2 represents the tension of CO_2, and pO_2 represents the tension of oxygen. Hydrogen ion concentration of blood serum is represented as $[H^+]$; r denotes the Donnan ratio between cells and serum, and v denotes the volume of the cells.

A very large number of related equations may be derived from the above set of five equations. As a whole, the set shows five relations among seven objects taken three at a time. The set therefore implies a large number of additional relationships. The total number of combinations of seven objects taken three at a time is given by:

$$\frac{7\times6\times5}{3\times2\times1} = 35 \text{ combinations}$$

Therefore there are 35 analogous equations showing the respiratory changes in blood. These may be represented by contour line charts which may then be transformed into d'Ocagne nomograms.

An equation of the form

$$CO_2 = f_1(pCO_2, pO_2)$$

may be arranged in two other equivalent forms:

$$pCO_2 = f'(CO_2, pO_2)$$

and

$$pO_2 = f''(CO_2, O_2)$$

The three relationships may be represented as sets of two-dimensional curves. Three such arrangements are possible for each of the 35 possible combinations of seven variables taken three at a time. Therefore the set of all possible arrangements among the seven variables or functions implies a total of $(7\times5\times3)$ or 105 two-dimensional Cartesian graphs. Thus in any biological system, the state of which depends on two independent variables, a very large number of two-dimensional diagrams are required to describe all the relationships; the number of possible combinations may become very high.

It then becomes necessary to transform the Cartesian contour line charts to nomograms of the d'Ocagne type. The method of making this transformation for growing and developing skeletal muscle will be illustrated in the following section.

Growth and Development of Rat Skeletal Muscle

The fundamental equations for electrolyte balance and electrical potentials in skeletal muscle are expressed as follows (chapters IV–VIII):

$$\Delta\mu_i = (\Delta\mu_i^\circ)' + RT\ln\frac{(c_i)''}{(c_i)'} \qquad (13.1)$$

$$\Delta\mu_{Na} = \Delta\mu_{Na}^\circ + RT\ln r_{Na} \qquad (13.2)$$

and

$$FE = FE^\circ + FE_{Na} \qquad (13.3)$$

The conversion factor from calories to millivolts is 1 kcal = 43.4 mV. Thus when $\Delta\mu = 2$ kcal, the value of E is − 86.8 mV (a value of the typical order of magnitude for mammalian skeletal muscle). A value of 3 kcal for $\Delta\mu_{Na}^\circ$ signifies a value of − 130.2 mV for E°; by changing the sign, a value of + 130.2 mV is obtained for E_a. The spike potential is then (130.2 − 86.8) or 43.4 mV. This corresponds to −1 kcal for RT ln r_{Na}, and to a value of 0.2 for r_{Na}.

According to equation 13.3, the relationship between E, E° and E_{Na} is linear. Similarly, there is a linear relation connecting $\Delta\mu$, $(\Delta\mu_i^\circ)'$ and RT ln $(c_i)''$ for any of the intramuscular ions. Therefore, the set of intracellular ions (Na, K, Cl) is readily described by a family of straight lines, when the properties of skeletal muscle are plotted graphically in relation to age and development. Calcium and magnesium ions may be included when there are available data; this results in a well-ordered set of five kinds of ions in relation to age [Joseph, 1971a]. Describing the results by Cartesian nomograms, the general formula is applied for any ion:

$$E = (E_i^\circ)' - \frac{RT}{z_i F} \ln (r_i) \tag{13.4}$$

At 37 °C, this becomes:

$$E = (E_i^\circ)' - 61.7 \log_{10} (r_i) \tag{13.5}$$

where E is the resting potential of the muscle, and where $(E_i^\circ)'$ is a characteristic constant for each kind of ion. It is related to $(\Delta\mu_i^\circ)'$ by the relation:

$$(E_i^\circ)' + 43.4 \frac{(\Delta\mu_i^\circ)'}{z_i} = 0 \tag{13.6}$$

where $(\Delta\mu_i^\circ)'$ is the apparent change of standard chemical potential for the given kind of ion. The constant 43.4 mV is the conversion factor from kcal to mV.

In table XXXIII (chapter VIII), results are given for c_{Na}'', $\Delta\mu_{Na}$, $\Delta\mu_{Na}^\circ$ and E for rat muscle at the following ages in days: 2, 4, 8, 16 and 32. Values of $(c_K)''$ and $(c_{Cl})''$ are also given. From the data, values of $-\log r_{Na}$, $-\log (r_K)$ and $-\log (r_{Cl})$ are computed. The values of $\Delta\mu_i$ are correlated by the formula:

$$\Delta\mu_{Na} = \Delta\mu_K = -\Delta\mu_{Cl} \tag{13.7}$$

By means of equations 13.5 and 13.6, values of $(E_i^\circ)'$ and $(\Delta\mu_i^\circ)'/z_i$ are computed at each age. This yields a family of five parallel straight lines that constitute a Cartesian nomogram describing the ion distributions (as logarithmic functions), apparent standard chemical potentials and resting potentials as functions of age and development (fig. 3). In each case, the resting potential E is represented by the position of the straight line. The value is given by the sum of the ordinate for any ion, $(E_i^\circ)'$ and the abscissa,

$$-\frac{61.7}{z_i} \log (r_i).$$

Thus in skeletal muscle, electrolyte distribution, chemical potentials and electrical potentials may be described as sets or subsets of properties which are univariant with respect to time. This implies well-ordered biunivocal relationships among all the properties. Thus when chemical morphology is treated as the primary property of muscle, a large number of secondary properties are correlated with each other and with age by one-to-one or biunivocal correspondence.

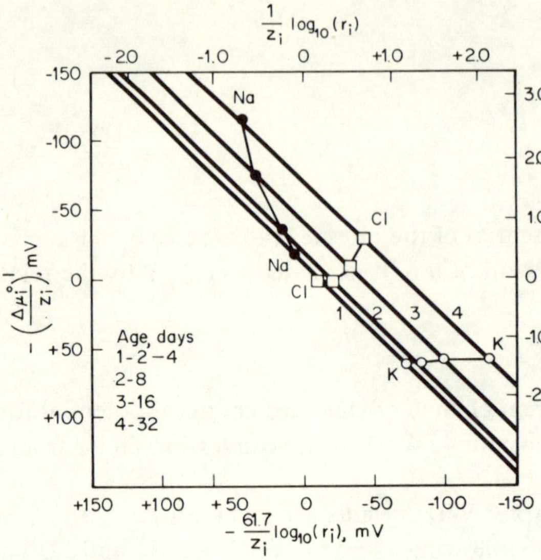

Fig. 3. Cartesian nomogram illustrating growth and development of rat skeletal muscle. Diagram shows standard distributions of the set of three ions (Na, K, Cl) at the following ages (in days): 2–4, 8, 16 and 32. Abscissae: distributions plotted as $-\log(r_i)$ and as $\frac{61.7}{z_i}\log(r_i)$. Ordinates: values of $\frac{(\Delta\mu_i^\circ)}{z_i}$ plotted as kcal and as mV.

D'Ocagne Nomograms

In a Cartesian nomogram, when the ordinates and abscissae are linear functions of the argument or independent variable (time, in the case of fig. 3), it is a simple matter to transform the results into a nomogram of the d'Ocagne type [Henderson, 1928].

In the case of skeletal muscle, the relationship between E, E_{Na} and E_a can be described by three parallel vertical lines (fig. 4). Then a straight line drawn between any two of the three corresponding points determines the position of the third. Since the values of $\Delta\mu_i$, r_{Na} and $\Delta\mu_{Na}^\circ$ are also implied by the above three values, many properties of skeletal muscle are implied by the d'Ocagne nomogram. The properties of the nomogram (fig. 4) are general for mammalian skeletal muscle. The specific properties are given by any two values on the generalized nomogram. These determine the third value; all three are functions of age, development and species.

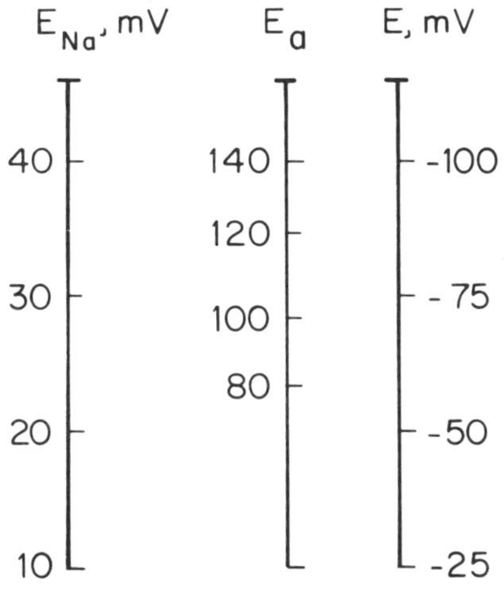

Fig. 4. D'Ocagne transformation of Cartesian nomogram shown in figure 3.

Normal and Pathological Blood

An example of a nomogram of normal adult human blood may be found in figure 158, page 200, of 'Blood' [HENDERSON, 1928]. This represents a number of properties of arterial and venous blood of CVC at rest. The properties are shown on a number of non-parallel vertical scales, and represent a set of ten variables which are coordinated by biunivocal relationships.

Any one of eight properties is determined by the values of oxygen tension (pO_2) and carbon dioxide tension (pCO_2). The horizontal lines represent the states of arterial and venous blood as determined respectively by the gas tensions in the alveoli of the lung and in the mixed venous blood of the right auricle. These lines transect the vertical scales at points which represent 8 other physicochemical functions which are determined by pO_2 and pCO_2 of the arterial or venous blood.

Ten vertical scales from left to right on the diagram are respectively: (1) the ratio of cell chloride to serum chloride, r_{Cl}; (2) the theoretical ratio, $r_{cl\,(theor.)}$; (3) the ratio of serum volume to whole volume, v_s/v_b; (4) base bound to bicarbonate, $BHCO_3$; (5) total CO_2; (6) CO_2 tension, pCO_2; (7) pH of cells; (8) pH of serum; (9) O_2 tension, pO_2; (10) oxyhemoglobin, HbO_2.

Blood alternates cyclically between the arterial and venous states; in the normal resting subject, it thus has one degree of freedom. The state is determined by two variables, pO_2 and pCO_2, but in the resting subject these are not independent – they are coordinated both in arterial and venous blood, as well as in lungs and body tissues.

In periods of exercise, the positions of the arterial and venous lines are altered. The displacements depend on the severity of the exercise. If the independent variables are displaced from pO_2, pCO_2, all other properties change from A, B, C . . . to A′, B′, C′ . . . The transitions are well-ordered; both the arterial and venous states are altered. The entire physicochemical system acquires an additional, or second, degree of freedom. Both sets, A, B, C . . . and A′, B′, C′ . . . are well-ordered and correlated by biunivocal correspondences. In the displaced state, as in the resting state, pO_2 and pCO_2 are not independent. Thus the displaced state depends on the specific values of pO_2' and pCO_2', which depend on the severity of the exercise for the given subject.

A pathological state (diabetic coma) is represented by figure 303, page 202, of 'Blood' [HENDERSON, 1928]. Here the entire nomogram is drastically displaced. The resting system acquires additional degrees of freedom, and all the relationships are distorted. In each case, the displacements would depend on the specific nature of the pathological state. Thus it would be impossible without a difficult detailed study to state the exact number of degrees of freedom by which the normal state is increased.

Pathological states such as diabetes or diabetic coma imply situations in which any specific property of any given cell or tissues (for example, $(c_K)''$) is determined by S, the set of all sets (M, N, O, P). Thus:

$$S = (M, N, O, P \ldots)$$

Therefore:

$$S = ((c_K)'', (c_{Na})'', (c_{Cl})'' \ldots)$$

where the sets of ionic concentrations are extended over all organs, cells and tissues. Thus a sufficiently severe pathological state must affect to a greater or less extent all the cells and tissues of the body. This implies an indefinite increase in the number of degrees of freedom of the whole organism, depending on the extent to which each kind of structure has been affected.

From what has been said, it is evident that any pathological state of sufficient severity connotes a widespread displacement of chemical morphology throughout the body. In any part of the physicochemical system, the displacement may possibly be represented by nomograms, alignment charts or mappings. These are based on changes of composition, chemical potentials and standard chemical potentials in all the affected regions. These may include well-ordered solid or semi-solid phases as well as the liquid phases of the *milieu intérieur*. The displacement of any phase or set of phases from the normal invariant state may be expressed by an increase of the number of degrees of freedom of the affected regions.

Chapter XIV: Beyond the Pointer Reading

'This signifies nothing because we do not know the nature of things.'
CLAUDE BERNARD
'. . . collection of a whole M of definite and separate objects of our intuition or our thought.'
GEORG CANTOR
'Mathematics is the science in which we do not know what we are talking about nor whether what we are saying is true or false.'
BERTRAND RUSSELL

We do not know the nature of things, and mathematics cannot supply the knowledge; hence we do not *know* the principles of mathematical biology. CANTOR's theory of sets and aggregates is based on 'definite and separate objects of our intuition or our thought'. It is therefore subjective and nominalistic. BERNARD's methods in general physiology and experimental medicine are designed to approach an understanding of things by the experimental method. The attempt to express this in terms of physical chemistry and mathematical symbolism requires the development of a hypothetical-deductive system of thought in which logical inferences can be tested and controlled by well-defined experiments.

This requires a conceptual scheme which permits no contradictions of the laws of nature. These determine the principles of biological causality. In the preceding chapters, an attempt has been made to develop such a conceptual scheme based on CARNOT's principle, GIBBS' 'Equilibrium of Heterogeneous Substances' and BERNARD's concept of an invariant *milieu intérieur*. CARNOT's principle might be stated in the form 'no work is performed by an invariant system'. The same thought was expressed by BERNARD [1878] in the form: 'We cannot therefore admit in living organisms a free vital principle which fights against external physical conditions.'

The application of thermodynamics to the study of homeostasis and physiological invariance suggests that BERNARD's ideas may be expressed as a principle of 'conservation of reversibility' in living organisms (chapter I). This enables us to develop a hypothetical-deductive system of mathematical reasoning which may be checked and verified by physical and chemical methods (pointer readings). On a day-to-day basis in mammalian physiology, irreversible processes are controlled by restrictions and constraints imposed by the first and second laws of thermodynamics. This accounts for the highly-ordered and well-regulated control of heat production, carbon dioxide production and the consumption of oxygen. Conservation of reversibility is shown by the day-to-day invariance of the mammalian organism. Control of the irreversible processes is brought about by order imposed by the invariant reversible processes.

Conservation of Reversibility

An example of this property is the approximately constant freezing point depression, Δ, of mammalian blood serum in all species at all ages [JOSEPH, 1971a, b]. The mean value may be estimated as about $-0.58\,^\circ C$; the extremes are about -0.56 and $-0.60\,^\circ C$. The value is related to the chemical potential of water, μ_{H_2O}, by the equation:

$$\mu_{H_2O} - \mu^\circ{}_{H_2O} = 5.26\,\Delta \tag{14.1}$$

where $\mu^\circ{}_{H_2O}$ is the chemical potential of pure water at the freezing point, $0\,^\circ C$. Then when Δ is $-0.58\,^\circ C$, μ_{H_2O} is -3.05 cal/mol. The constant 5.26 cal/deg. is the entropy of fusion of ice. The chemical potential of water in blood serum, $\mu'{}_{H_2O}$, is related to the chemical potential in somatic cells and tissues, $\mu''{}_{H_2O}$, by GIBBS' equation 77. Thus:

$$\mu'{}_{H_2O} = \mu''{}_{H_2O} \tag{14.2}$$

For the binary and ternary electrolytes, NaCl, KCl, $CaCl_2$ and $MgCl_2$, a form of GIBBS' equation 78 is employed:

$$\Sigma a_1 \mu_1 = 0 \tag{14.3}$$

At constant temperature and pressure, the chemical potential of water is related to the chemical potentials of the ions by the formula:

$$d\mu_{H_2O} + 0.018\,\Sigma c_i d\mu_i = 0 \tag{14.4}$$

This is a special form of GIBBS' equation 97, where 0.018 denotes the molecular weight of water divided by 1,000. Finally, the reversible electromotive force E is related to the change of chemical potential of any ion by the formula:

$$FE + \delta = 0 \tag{14.5}$$

where F is the Faraday constant, and where δ is the equivalent change of chemical potential of each of the five physiological ions. Equation 14.5 is a form of GIBBS' equations 687 and 688. It connotes that the five ions behave as a well-ordered set, which may be invariant or univariant in time.

Equations 14.2–14.5 are the conditions for reversible transport and balance of water and the five ions between any two phases of a well-ordered resting biological system. For any ion:

$$\Delta\mu_i = \Delta\mu_i^\circ + RT \ln \frac{c_i''f_i''}{c_i'f_i'} \tag{14.6}$$

where f_i'' and f_i' denote the ionic activity coefficients. By an appropriate definition of the standard state of the two phases, the coefficients are assigned the values of 1.0. Biological phases approach states of invariance at each resting state on the normal growth curve. Hence they do not behave as true solutions in which the values of f are obtained by extrapolation of experimental data to infinite dilution. The conventions that are applicable to ordinary salt solutions cannot be applied to cells and tissues. In these, the only possible reference state is the standard resting state, where f of all ions is assigned the value of 1.0 [JOSEPH, 1971a, b]. In mammalian cells and tissues:

$$\Delta\mu_i = \Delta\mu_i^\circ + RT \ln \frac{c_i''}{c_i'} \tag{14.7}$$

where $\Delta\mu_i^\circ$ is the change of standard chemical potential. In the Born theory of ionic hydration energy [BORN, 1920]:

$$\mu_i^\circ = \frac{N_i z_i^2 \varepsilon^2}{2\,D\,b_i} \tag{14.8}$$

where N_i is Avogadro's number, z_i is the ionic charge and ε is the electrostatic unit of charge. D is the dielectric constant, and b_i is the corrected ionic radius [LAIDLER and PEGIS, 1957].

Between two phases of the same dielectric constant, $\Delta\mu_i^\circ$ is zero. This is the case in the paradigmatic model of the classical Donnan equilibrium [DONNAN, 1911]. This represents two aqueous phases separated by a membrane that is permeable to water and inorganic ions, but which is impermeable to proteins and other macromolecular polyelectrolytes. In this model, DONNAN did not take into account possible changes of the aqueous dispersion medium between the two phases. Consequently, the model cannot represent a heterogeneous system in which water coexists in two or more immiscible states. Therefore the values of $\Delta\mu_i^\circ$ do not appear in DONNAN's formulae for 'membrane equilibrium' or 'membrane potentials'. The theoretical basis for these formulae is not adequate to account for water and electrolyte balance or for resting or action potentials in biological systems.

From this, many authors have concluded that the 'Gibbs-Donnan' equilibrium cannot be applied to living cells and tissues. On the basis of this and other negative premises (failure of the classical Nernst equation for electromotive force), they have concluded that there are 'sodium pumps' and other mechanisms of active transport in metabolizing cells. Reasoning from this negative premise, it has been postulated that the resting electromotive force is irreversible, and that it depends on selective ion permeabilities and active transport of ions.

In thermodynamics, the only correct general formula for a stable reversible electromotive force is given by equation 14.5 (GIBBS' equations 687 and 688). The condition for a stable invariant system is thereby satisfied, namely that the reversible work in an isothermal system is zero. This is CARNOT's condition for invariance. It is also a necessary condition required for conservation of reversibility. Equation 14.5 can be expressed in the form:

$$\mathrm{FE} + \frac{\Delta\mu_i}{z_i} = 0 \tag{14.9}$$

where $\Delta\mu_i/z_i$ or δ is the same for each of the ions. This is a condition for reversibility, invariance and for the stability of a well-ordered set of five ion concentrations and potentials. Equation 14.9 can be considered to represent the limiting formula for any phase boundary potential:

$$\mathrm{FE} + \int \Sigma \frac{t_i}{z_i} d\mu_i = 0 \tag{14.10}$$

where t_i denotes the transport number of a given ion. The condition for reversibility requires that in the limit:

$$\int \Sigma \frac{t_i}{z_i} d\mu_i = \delta \tag{14.11}$$

This yields equation 14.9. Thus the true condition for stability, invariance and reversibility requires that E be independent of transport numbers and permeabilities.

The equations for 'membrane potentials' that have been developed in the theory of 'active transport' do not satisfy the conditions for reversibility. They represent irreversible electromotive force, which depends on the properties of the membrane rather than on the states of intracellular and extracellular phases. This is also true of the treatment of action potentials in the theory of 'active transport'. These have been considered to depend on changes of membrane permeabilities during production of the action potential. Repolarization has been considered to depend on the action of the 'sodium pump'.

All these conclusions depend on the fundamental assumptions: (1) uniform dielectric properties in a system in which water exists in only one state of aggregation; (2) maintenance of thermodynamically unstable states by utilization of metabolic energy to maintain ion 'concentration gradients' across selectively permeable membranes. These and other *ad hoc* assumptions have been found necessary to account for experimental observations which could not otherwise be explained by the 'Donnan equilibrium'.

This line of reasoning is entirely vitiated when it is perceived that GIBBS' 'equilibrium of heterogeneous substances' is an entirely different paradigm from that given by DONNAN's 'theory of membrane equilibrium and membrane potential in the presence of a non-dialyzable electrolyte'. It is clear from the titles of these works that DONNAN is dealing with a limiting case of GIBBS' general treatment. It is a case of equilibrium in which both phases are aqueous solutions, i.e. phases that are homogeneous and that show perfect mixing of the various components. GIBBS deals with the most general case of systems in which the phases are totally or partially immiscible, and in which dielectric properties may be considered to vary throughout the system. In biological systems where water may exist in more than one state of aggregation, DONNAN's theory of 'membrane equilibrium' can represent only a theoretical limiting case. Hence it is erroneous to think of the 'Gibbs-Donnan' equilibrium as representing the general conditions of heterogeneous equilibrium. The fundamental negative premise of the theory of 'active transport' cannot be granted. In any heterogeneous system coexistent phases must be expected to show stable concentration gradients of all the components, including all the inorganic physiological ions.

In GIBBS' theory, electrolyte balance and reversible electromotive force are described by four equations: 14.2–14.5. Standard chemical potentials and dielectric constant are interrelated by the Born theory [BORN, 1920]. An additional condition is required to meet the phase rule conditions for invariance [JOSEPH, 1971a, b]:

$$x + \Sigma(c_i)''z_i = 0 \tag{14.12}$$

This describes the relation between colloidal charge x and ion concentrations. It represents the condition for acid-base balance. Like the dielectric properties, acid-base balance depends on morphology of the cell rather than on its metabolism. The theory of 'active transport' implies that the 'concentration gradients' are maintained by metabolic energy. This is based on the idea that morphological structure is irrelevant; the intracellular cytoplasm is treated as a dilute salt solution. The idea that cells spend energy to maintain a steady state contradicts CARNOT's principle that the reversible work of any process in an isothermal, invariant system is zero [JOSEPH, 1966a]. Thus active transport is incompatible with conservation of reversibility.

Logical Principles

Biological systems are restricted open systems [JOSEPH, 1971b]. They maintain permanent structures and forms of internal energy. Homeostasis implies a number of fixed internal and external boundaries and fixed conditions of state. Organisms are also open systems, however, because they require external sources of oxygen and nutrients as well as outlets for expired carbon dioxide and heat. They are also open with respect to limited numbers and kinds of nutrients, and are restricted with respect to inedible or toxic substances.

The situation is illustrated by figure 5, which shows two overlapping circles. The first of these, I, represents the class of all open systems. Circle II represents the class of all closed systems. The shaded area III represents all restricted open systems, i.e. those that are closed in one or more respects. Then we have the proposition: all living systems are in III – this must always be in one universe of discourse. No organisms are open in all respects nor closed in all respects. The closed part of III represents the fixed parts of the morphology of cells and tissues. The open part represents respiratory gases, metabolites, nutrients and heat, which may be in permanent states of flux. However, in the standard state, all rates of transport and exchange would be determined by the constant standard state of respiratory metabolism. This represents a state of perfect energy and electrolyte balance, which is controlled by restrictions and constraints imposed by the first and second laws of thermodynamics. This accounts for the highly-ordered and

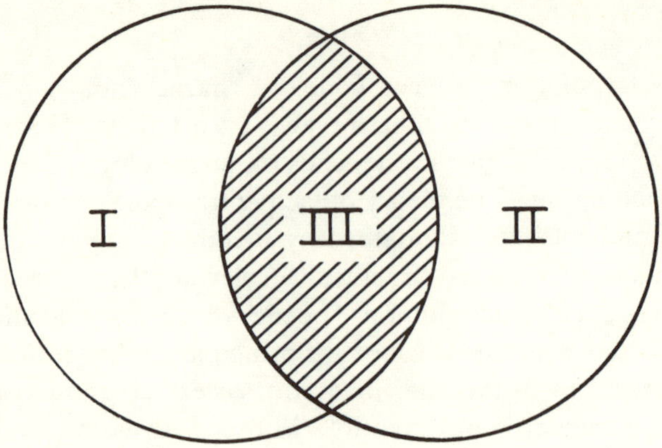

Fig. 5. Physicochemical conditions for existence of stable, restricted open systems. I: set of all open systems; II: set of all closed systems; III: set of all restricted open systems. All living cells and tissues are in set III. This is also a condition for physiological invariance (fig. 2).

well-regulated control of the irreversible processes such as production of heat and carbon dioxide and the consumption of oxygen. Conservation of reversibility is shown by the day-to-day invariance of the mammalian organism. Regulation and control of the irreversible processes of nutrition and respiration are also provided by the laws of thermodynamics. Control of the irreversible processes is brought about by order that is imposed by the invariant reversible processes.

Another fundamental logical problem is illustrated in figure 6, which shows two different ways of relating morphology, metabolism and transport. Three concentric circles are labelled M (morphology), m (metabolism), and t (transport). According to thermodynamics, m is determined by temperature and chemical morphology. Therefore:

$M \supset m$

In the invariant standard state, all transport processes are determined by the morphology, M:

$M \supset m \supset t$

This is shown in figure 6, I, which arranges the circles in the order M, m, t, indicating that both m and t are determined by chemical morphology. Thus the respiration of nerve differs from that of skeletal muscle because of morphological differences.

In the theory of active transport, the ion distributions are considered to depend on m rather than on M. Accordingly, the fundamental assumption, expressed logically, is that:

$$m \supset t \supset M$$

Thus chemical morphology, including electrolyte composition, is considered to be determined by active transport, t, which requires a supply of metabolic energy, m. This is represented by figure 6, II, in which the outer circle is labelled m, and the inner circle is labelled M. This is in direct contradiction to the thermodynamic point of view, in which chemical morphology and physicochemical state determine all secondary properties, including metabolism and transport [JOSEPH, 1971a, b].

The thermodynamic theory describes the entire system by an invariant or univariant set of well-ordered functions which can be described by Cartesian or d'Ocagne nomograms (chapter XIII). The theory of active transport, based on irreversible mechanistic properties, implies a large number of uncoordinated *ad hoc* mechanisms (pumps, carriers and selective permeabilities), to which the concept of degrees of freedom cannot be applied. Therefore, the principle of causality does not rest on the laws of thermodynamics, but appears to rest on arbitrary subjective and nominal-

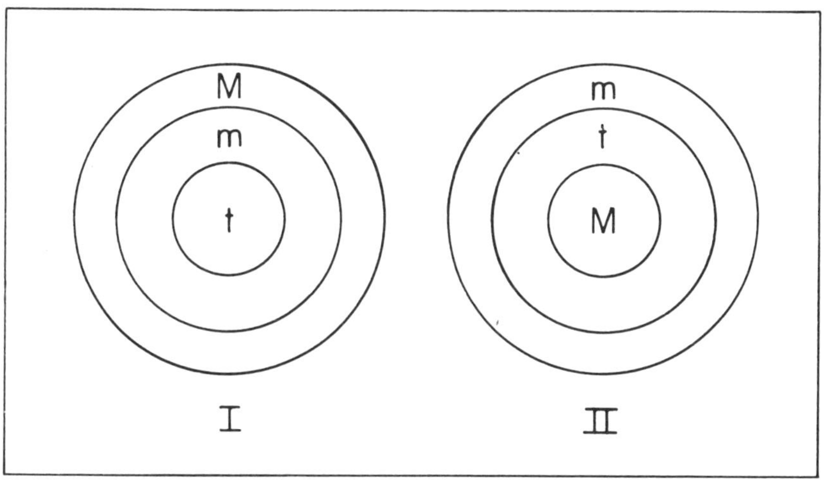

Fig. 6. Conceptual schemes for relations between chemical morphology (M), respiratory metabolism (m) and irreversible transport of CO_2, oxygen, nutrients and heat. I: thermodynamic; II: mechanistic (active transport).

istic principles. By eliminating chemical morphology from the conceptual scheme, the theory of active transport also excludes the possibility of *changes* of configurational entropy or of dielectric energy as fundamental processes (chapter IX). This seems also to eliminate rational explanations of irritability or of contractility, which can be readily explained by changes of chemical morphology and by changes of state of water and electrolytes.

Pointer Readings

The title of this chapter is suggested by chapter XII of 'The Nature of the Physical World' by EDDINGTON [1929]. As the exact sciences, physics and chemistry, have come to depend more and more on the use of accurate measuring devices, their progress depends on the triple correspondence: (a) the image formed in the human mind but which is not in the external world; (b) an inscrutable counterpart in the external world; (c) a set of pointer readings determined by exact science and related to other sets of readings.

EDDINGTON makes clear that the physicist deals with concepts and words such as length, angle, force, and so forth, which are called 'physical quantities'. In the course of time, all such quantities have tended to lose any metaphysical or non-operational character originally assigned to them. In modern physics all such quantities are carefully defined in accordance with the exact methods used in their measurement. According to BRIDGMAN [1927], all significant definitions and measurements must be 'operational'. Non-operational concepts, quantities and magnitudes are often found to be 'meaningless'. To quote EDDINGTON: 'EINSTEIN's theory makes a clean sweep of those pious opinions and insists that each physical quantity should be defined as the result of certain operations of measurement and calculation.' By 'pious opinions', EDDINGTON has in mind the non-operational definitions which lead to meaningless concepts. It should be understood that these criteria for determining the physical significance of exact 'pointer readings' are now accepted by chemists as well as by physicists.

In the 20th century, the biomedical sciences have adapted many of the exact methods of physics and chemistry to physiology and medicine. These include the techniques of roentgenology and electrocardiography, as well as the exact potentiometric techniques of measuring resting electrical potentials and action potentials of nerve or muscle preparations. The pointer readings are also used in exact chemical methods which involve photometry, colorimetry or various kinds of spectroscopy. If physiologists are not to misuse and misapply these techniques, it is necessary that the conceptual schemes of their science must be freed of non-operational, meaningless concepts that are not rigorously defined.

In the study of muscular contraction, for example, the directly observed 'physical quantities' are resting potentials, action potentials, work, tension, length, fiber radii, and other magnitudes which are physically measurable. Operationally defined chemical magnitudes are composition of water, electrolytes and proteins. Organic phosphates such as ATP and creatine phosphate may also be chemically determined and operationally defined. In the conceptual scheme that has been developed in previous chapters, chemical morphology is described by a set of chemical components that includes water, electrolytes, macromolecular polyelectrolytes, and other sets of intracellular substances. Physical properties such as electrical potentials, tension, dielectric energy and maximal work comprise a secondary set of properties which is determined by chemical morphology treated as the primary set (chapters IX–XIII). This requires a conceptual scheme in which the two sets are related by biunivocal correspondences based on the theory of heterogeneous equilibrium.

Other conceptual schemes in which transport, permeability and metabolism are taken as the primary set of properties may be shown to be meaningless on the basis of the following criteria:

1. Such schemes do not satisfy the criteria for energy balance.
2. They do not account quantitatively for chemical composition and electrical potentials of cells and tissues.
3. They do not account for tension and work of skeletal muscle.
4. They are based on non-operational concepts of permeabilities and transport mechanisms (pumps and carriers).
5. They fail to yield biunivocal correspondences between the sets of primary and secondary properties, as studied by the pointer readings.
6. They do not consider such thermodynamic properties as configurational entropy, free energy and dielectric energy.
7. They are based on a uni-state theory of water structure, and do not consider the lability of water and electrolytes in more than one intracellular state of aggregation.

Thus the external world of animal cells and tissues remains inscrutable in spite of the accuracy of the various kinds of pointer readings studied by the physiologist. This might have been anticipated, since as FISCHER and HOOKER [1933] pointed out, the conceptual scheme of the physiologists considered extracellular and intracellular fluids only as dilute salt solutions within semi-permeable membranes. No distinction was made between GIBBS' 'equilibrium of heterogeneous substances' and DONNAN's 'theory of membrane equilibrium and membrane potentials'. The 'Donnan equilibrium' is quite inadequate to establish the thermodynamic correspondences between the primary and secondary properties of cells and tissues. It therefore is insufficient to correlate the observed pointer readings with the otherwise inscrutable nature of living cells and tissues.

Morphology and Metabolism

In GIBBS' theory of heterogeneous equilibrium, the distribution of chemical substances is determined by the thermodynamic properties of the various components and phases. Stability, invariance and conditions of constraint are determined by two conditions:

$$\Delta G_{rev.} = 0 \qquad (14.13a)$$

$$\Delta G_{irrev.} < 0 \qquad (14.13b)$$

Equation 14.13a is the condition for reversible transport of electrolytes and water in an invariant, isothermal system. It applies to reversible processes in steady states as well as to systems at equilibrium. The inequality (equation 14.13b) expresses one of the conditions for invariance in the same system, i.e. that oxygen, carbon dioxide and nutrients be transported irreversibly. Maintenance of invariance or homeostasis requires that the various rates of irreversible transport must be highly coordinated with the rates of irreversible production of carbon dioxide and heat, and with the irreversible uptake of oxygen, glucose, lipids and other nutrients.

Each type of cellular structure is characterized by its own normal basal metabolic rate. Hence it is logical to write:

$$M \supset m \supset t$$

where M denotes chemical morphology, m denotes basal metabolic rate, and t signifies the standard rate or irreversible transport. The condition for an invariant state of homeostasis requires that:

$$m = t$$

The situation is described by the system of concentric circles shown in figure 6, I. The symbol M represents chemical morphology, which implies both chemical composition and physicochemical state. Skeletal muscle, which along with heart, has a very high rate of metabolism and irreversible transport, may change its physicochemical state very rapidly with respect to tension and length. Since this process may occur within a few milliseconds, it involves no necessary change of chemical composition. Therefore, when M changes to M′, this necessarily implies a physicochemical change of state, but does not necessarily imply a change of water, electrolyte or macromolecular composition. The converse is not true. A change of composition always implies a change of physicochemical state and of thermodynamic functions.

Since $M \supset m$, any change of morphology implies a change of respiratory metabolism, which is determined to a great extent by the rates of oxidation of glucose and lipids. When M is in its standard state, m is also characterized by a standard invariant value. In the isotonic contraction of muscle, external work may be performed at the expense of dielectric energy (chapter IX). This involves simultaneous changes of state of the contractile fibers, water and electrolytes. Water passes from a state of low dielectric constant ($D'' = 30$) to a state of high dielectric constant ($D'' = 80$). This implies an increase of entropy or disorder, and a change in its character as a solvent. Anaerobic glycolysis is stimulated, and mitochondrial oxidation of lipids is inhibited. This implies an increase of the respiratory quotient and an increase of the metabolic rate and heat production. Symbolically this is represented as:

$$M' \supset m' \supset t'$$

A steady state requires that $m' = t'$. Otherwise the muscle falls into oxygen debt and accumulates CO_2, lactic acid and inorganic phosphate. If rates of heat transport are inadequate, the temperature rises. The increase of inorganic phosphate is largely at the expense of creatine phosphate.

Thus in the thermodynamic conceptual scheme, represented by figure 6, I, $M \supset m$, and $M' \supset m'$. The work and free energy of contraction are represented by changes of configurational entropy and dielectric energy. These are maintained at standard values by the standard rates of glycolysis characteristic of the chemical morphology of the given kind of muscle fiber. Repolarization and recovery of a contracted fiber are brought about by an increased rate of glycolytic metabolism.

In the alternative scheme of metabolism, transport and morphology that is shown in figure 6, II, the relationships are given by:

$m \supset t \supset M$

and

$m' \supset t' \supset M'$

Thus the contractile process is represented by the change from M to M′; this depends fundamentally on the change of metabolism, m to m′. The conditions for reversible and irreversible transport, equations 14.13a and 14.13b are thermodynamic and are not considered. The reasoning is therefore mechanistic. Such schemes omit the consideration of dielectric energy, configurational entropy, and free energy. The energy of contraction is generally considered to be derived from the breakdown of ATP to ADP. The reasoning follows logically from the mechanistic conceptual scheme (active transport). However, the fundamental premises cannot be granted, because the concepts of mechanistic pumps and carriers have been shown in an earlier section to be non-operational, and therefore meaningless. They also have been shown to be based on DONNAN's uni-state theory of water in heterogeneous systems rather than on GIBBS' very general 'equilibrium of heterogeneous substances'.

Animal Behavior and Muscular Contraction

'By no means every physico-chemical process represents a life process, though every event in a living organism is a physicochemical phenomenon' [LORENZ, 1970]. Thus the behavior of animals, depending on the coordinated action of brain, afferent, efferent and motor nerves on skeletal muscle, may be regarded as a well-ordered pattern of physicochemical phenomena. If it is true that ATP is the universal source of energy for muscular contraction, its breakdown must be the unique cause of contraction. Logic would then require that the integrative action of the motor functions implies direct and coordinated intracellular responses of ATP to external stimuli and to voluntary control exerted by the higher centers of the cerebral cortex. This implies direct rather than indirect control of cellular metabolism by the central and autonomic nervous systems. The ATP hypothesis of muscular contraction thus requires testing from the point of view of the science of ethology or animal behavior.

The first point to be considered is the known chemical composition of skeletal muscle with respect to ATP. Analysis shows that the intrafibrillar composition is of the order of 5 mM/kg water. The bond energy yielded by the breakdown of ATP is of the order of 8 kcal/mol. The quantitative breakdown of 5 mM would then yield a maximum value of 40 cal/kg water. This is about half the value of $c_{Na}''\Delta\mu_{Na}^{\circ}$, the dielectric energy. According to the results of chapter IX, the maximal work of the muscles of the human arm or leg corresponds to the dielectric energy rather than to the bond energy of ATP.

When the concentration of ATP is expressed as g/kg water, the value is found to be of the order of 1 g or 0.1% by weight. ATP is then to be regarded as a trace substance rather than as a bulk component. On the other hand, the bulk components, water, proteins and electrolytes, constitute nearly all the mass (96–98%) of skeletal muscle. This mass is directly or indirectly connected to afferent, efferent and motor fibers of the central nervous system, which integrate the behavior of the entire neuromuscular system. This system of controls involves highly intricate behavior patterns that govern all the motor phenomena under natural conditions of life [LORENZ, 1970, 1971].

According to what has been said previously, the conceptual scheme that relates morphology to metabolism may be symbolized in the form:

$$N \supset M \supset m$$
$$N' \supset M' \supset m'$$

where N denotes nervous control in the resting or unstimulated state, and where N′ denotes a stimulated state. Thus m, the resting metabolism, is determined by M, the chemical morphology, as defined by chemical composition and physicochemical state. It follows that m′, the stimulated state of respiratory metabolism, depends on the change of physicochemical state of the muscle from M to M′; this is caused by the change of nervous stimulation from N to N′.

According to this view, the function of ATP is to react as a phosphorylating agent in the accelerated rate of anaerobic glycolysis (chapter XII). This function is stimulated by the change of chemical morphology from M to M′. Metabolism is then regulated by physicochemical state. This is controlled by action of the central nervous system and by mechanical forces and tensions exerted by neighboring muscular and connective tissues.

This concept is in harmony with BERNARD's views as to the function of the *milieu intérieur*. This function is to hold all body fluids in constant states, which protect all the somatic cells and tissues from change occurring in the external environment. Thus an intracellular component such as ATP could not possibly be influenced directly by sudden auditory or visual stimuli to which animals in a state of nature must instantly respond. These stimuli act on the animal organism through a causal sequence in which the first step is stimulation of some of the sensory nerves, especially of the eye or ear. The final step in the sequence is the motor response, which leads to an altered metabolic rate.

Far down the chain of causality is the metabolic response, involving glycolysis, as mediated by ATP and other organic phosphates. The immediate source of energy for the motor response is the supply of configurational free energy or configurational entropy that is maintained in the unstimulated state of the myofibrils. It is characteristic of the metabolic (or ATP) theory of contraction that energy transfer involves only chemical bond energy. In the thermodynamic theory, the requirement is met by a transfer of dielectric energy or configurational entropy (chapter IX).

In the latter theory, muscle fibers transfer dielectric energy by reason of the mechanical or tensile forces which they exert on each other. Thus the contraction of any given fiber or group of fibers is never independent of that of other fibers. It also depends on the mechanical forces exerted by tendon, cartilage, and other kinds of connective tissue. Since metabolism is a function of chemical morphology and physicochemical state, the metabolism of any fiber or set of fibers depends on the mechanical forces exerted by other fibers or sets. Thus the entire system of fibers, sets of fibers, and groups of synergistic or antagonistic muscles is integrated not only by the nervous system but also by the mechanical and tensile forces which are in constant states of flux during periods of muscular activity. These forces and tensions depend on anatomical structure, chemical morphology and physicochemical state of the entire neuromuscular system. Morphology and state are what an athlete develops in the course of his training. The intracellular respiratory metabolism develops concomitantly with chemical morphology. This is the primary property that determines all the secondary properties of the organism.

Beyond the Pointer Reading

A biological *process* is described by a change of state which, depending on its complexity, involves a larger or smaller number of simultaneous changes in the sets and subsets that comprise the system. For example, in a simple voluntary movement of one finger, the neuromuscular changes can be described by the process $S \rightarrow S'$, where S is the set (N, M, m) that describes the state of nerve (N), morphology of muscle (M) and metabolism of muscle (m).

Thus even in this trivial muscular movement, three coordinated sets of processes are involved, coordinating the flexor and extensor muscles and all their fibers, the motor nerves, and sets of metabolic processes in each kind of cellular structure. Voluntary changes of metabolism cannot be produced independently of muscular contraction. Therefore the concept of metabolic control of electrolyte balance is non-operational.

The muscular activity of a concert pianist in performance is, of course, much more complex. All the muscles of the arms, hands, fingers and wrists are brought into play. Musical expression requires delicate balance of rhythm, 'touch', tone and intensity (pianissimo to fortissimo), under the most sensitive neuromuscular control. The use of the pedals, controlled by the muscles of the legs, ankles and feet, is said to be an art in itself. The musical conception is expressed by higher centers of the brain. The pianist expresses his own personality, developed through years of study, practice and training of fingers, arms and wrists. Thereby, he has educated his neuromuscular system, coordinated it with his brain, and brought all this behavior into communication with his *musical* faculties – the expression of Beethoven, Brahms, Chopin or Liszt. In the great pianists, musical faculties are always combined with intellectual faculties: the understanding of audiences, critics, and their levels of culture and intelligence.

In the case of a great orchestra, playing under a great conductor, the complexity is magnified. A concert performance involves the brain and neuromuscular systems of a hundred sensitive and skilled artists. The set of instruments comprises a number of subsets: strings, woodwind, brass and percussion. Tempo, rhythm, tone and expression are coordinated by the movements and facial expressions of the conductor. The *musical* expression of a Beethoven symphony results from the conductor's education and experience; these are related to musical tradition and the experience of hundreds of orchestras, dating from the first performance of the work. A physiologist may conceivably study a whole performance as a complex set of simultaneous and synchronized movements of a hundred musicians. A student of muscular metabolism may regard it as the summation of millions of intracellular phosphorylations. It would be impossible to convince the audience or the musicians that the whole experience can be reduced to the cellular level. It is a mistake, then, to attempt to reduce all biology to cellular biology.

A fundamental fallacy of the biological sciences may then be the tendency to reduce everything in nature to mechanism: simple location and motion of particles in space and time. This tendency dates historically from the 17th century – the period of Descartes and Newton. But it is older. The Greeks knew it in the atomic doctrines of Democritus.

In biology, the corrective would seem to be in studying nature as a hierarchy of sets, governed by universal laws and processes. Thus animal behavior is to be understood in relation to the place in nature of a given species. Physical biology and ecology study the set of all species, animal and plant. But they must also study the set of all habitats – the environment. Any animal is a set of all its cells and tissues. To describe behavior as beginning at the cellular level finally results in reducing the art of the pianist or violinist to his enzymes.

WHITEHEAD and other speculative thinkers have long studied the nature of such fallacies. Perhaps they are the inevitable result of the natural human tendency to generalize a limited experience beyond its legitimate boundaries. Mechanist thinkers are especially susceptible to the fallacy of misplaced concreteness. In one of his books, 'Process and Reality', WHITEHEAD [1929] has explained his attempts to find the corrections for simple location and for Cartesian dualism. In biology a similar dualism is found in the apparent antagonism between morphology (M) and metabolism (m). A dualistic view is compelled to choose between two alternative schemes:

$M \supset m$, and $m \supset M$

At the present time, cellular biologists have chosen the latter course; the reason is that chemical morphology has been treated as a secondary rather than as a primary property. Hence, the study of morphological growth, development and aging has fallen into disfavor, and primary attention has been directed toward intermediary metabolism. The dualistic balance has been weighted in favor of intracellular respiratory chemistry.

WHITEHEAD's concept of *process* as the fundamental reality would seem to restore the balance. This could be expressed in the form:

$$P \supset (M, m) \qquad \text{(fig. 7)}$$

Thus process, P, implies effects on the set of biological subsets embraced by M (chemical morphology and state) and m (respiratory metabolism). This concept tends to correct the fallacy that results from the elimination of morphology as a primary concept, and its substitution by metabolic transport. This fallacy, as has been pointed out in earlier sections, is a result of the idea that water exists everywhere in the organism in one state of

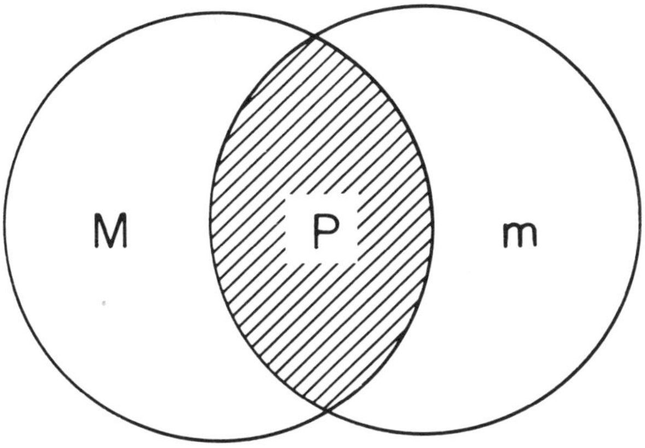

Fig. 7. Physicochemical conditions for coordination of morphological and metabolic processes in living organisms. M: set of all morphological states in living and non-living systems; m: set of all metabolic states in living and non-living systems. All P is in both M and m. M and m depend on P, and cannot vary independently of P, the biological process in a living system of cells and tissues.

aggregation – the liquid state. The recognition that water exists in other labile states of aggregation opens the way to understanding of irritability, contractility and other kinds of behavioral responses that may be denoted as biological processes. Individual processes, in living mammals, are compounded into the very complex behavior patterns that describe the actual responses at the social and ecological level. The individual processes always result in changes of physicochemical state and respiratory metabolism at cellular and submicroscopic levels.

References

Adams, H.: The tendency of history (Macmillan, New York 1919).

Adrian, R. H.: The effect of internal and external potassium concentration on the membrane potential of frog muscle. J. Physiol., Lond. *133:* 631–658 (1956).

Atwater, W. O.; Woods, C. D., and Benedict, F. G.: US Dept. Agr. Office Exp. Stn, Bull. *44:* 51 (1897).

Benedict, F. G. and Cathcart, C. I.: Muscular work (Carnegie Institution, Washington, D.C. 1913).

Berendsen, H. J. C.: Biological significance of water structure; in Person Biology of the mouth, publ. No. 89 (Amer. Ass. Adv. Sci., Washington, D.C. 1968).

Bernal, J. D.: The structure of water and its biological implications. Symp. Soc. exp. Biol. *19:* 17–32 (1965).

Bernard, C.: Leçons sur les phénomènes de la vie, communs aux animaux et aux végétaux (Ballière, Paris 1878).

Bernard, C.: The cahier rouge (1856) (Schenkman, Cambridge, Mass. 1967).

Boltzmann, L.: (Populäre Schriften, Leipzig 1905).

Bondareff, W.: Morphology of connective tissue ground substance with particular regard to fibrillogenesis and aging. Gerontologia *1:* 222–233 (1957).

Born, M.: Volumes and heats of hydration of ions. Z. Phys. *1:* 45–48 (1920).

Bridgman, P. W.: The logic of modern physics (Macmillan, New York 1927).

Brictaux-Grégoire, S.; Duchateaux-Bosson, G.; Jeauniaux, G. et Florkin, M.: Constituants osmotiques actifs des muscles du crabbe chinois *Eriocheir sinensis* adapté à l'eau douce ou à l'eau de la mer. Arch. int. Physiol. Biochim. *70:* 273 – 286 (1962).

Brooks, W. H.: The foundations of zoology (Macmillan, New York 1899).

Bungenberg de Jong, H. G.: in Kruyt, Colloidal science, vol. 2, chap. 8–11 (Elsevier, New York 1949).

Bütschli, G.: Untersuchungen über microscopische Schaume und das Protoplasma (Leipzig 1892).

Cannon, W. B.: The wisdom of the body (Norton, New York 1932).

Cantor, G.: Contributions to the founding of the theory of transfinite numbers. I. Math. Ann. *46:* 481–512 (1895).

Cantor, G.: Contributions to the founding of the theory of transfinite numbers. II. Math. Ann. *49:* 207–246 (1897).

Casella, C.: Tensile force in total striated muscle, isolated fibers and sarcolemma. Acta physiol. scand. *21:* 380–410 (1950).

Chan, D. K. O.; Chester Jones, I.; Henderson, I. W., and Rankin, J. C.: Studies on the experimental alteration of water and electrolyte composition of the eel *(Anguilla anguilla,* L). J. Endocrinol. *37:* 217–317 (1967).

CHRISTENSEN, J. J.; HILL, J. O., and IZATT, R. M.: Ion binding by synthetic macrocyclic compounds. Science *174:* 459–467 (1971).

COHN, E. J.: The physical chemistry of the proteins. Physiol. Rev. *5:* 349–437 (1925).

DARROW, D. C.; HARRISON, E. E., and TAFFEL, M.: Tissue electrolytes in adrenal insufficiency. J. biol. Chem. *130:* 487–502 (1939).

DARWIN, C. G.: The next million years (Doubleday, Garden Vity, N.Y. 1953).

DEBYE, P.: Polar molecules (Chem. Catalog Co., New York 1929).

DICKERSON, J. W. T. and WIDDOWSON, E. M.: Chemical change in skeletal muscle during development. Biochem. J. *74:* 247–257 (1960).

DONNAN, F. G.: Theory of membrane equilibrium and membrane potential in the presence of a non-dialyzable electrolyte. Z. Elektrochem. *17:* 572–581 (1911).

EAGLE, H.: Metabolic controls in cultured mammalian cells. Science *148:* 42–51 (1965).

EDDINGTON, A. S.: The nature of the physical world (Macmillan, New York 1929).

EINSTEIN, A.: Investigations on the theory of the Brownian movement (1905–1908) (Methuen, London 1926).

FENN, W. O.: Electrolytes in muscle. Physiol. Rev. *16:* 450–487 (1936).

FISCHER, M. H. and HOOKER, M. O.: The lyophilic colloids (Thomas, Springfield 1933).

FREGE, G.: What is a function? (1904). Translations from the philosophical writings of Gottlob Frege (Philosophical Library, New York 1952).

FREY-WYSSLING, A.: Submicroscopic morphology of protoplasm (Elsevier, Amsterdam 1953).

FRITZ, I. B.: Factors influencing the rates of long-chain fatty acid oxidation and synthesis in mammalian systems. Physiol. Rev. *41:* 52–129 (1961).

FUDEL-OSIPOVA, S. I. and MARTYNENKO, O. A.: Early ontogenetic development of membrane potential of muscle fibers in rats. Fed. Proc. Transl. Suppl. *T28–31* (1964).

GERSH, I. and CATCHPOLE, H. R.: The nature of ground substance of connective tissue. Perspect. Biol. Med. *3:* 292–349 (1960).

GIBBS, J. W.: On the equilibrium of heterogeneous substances (1875); in Collected works, vol. 1 (Longmans, Green, New York 1928).

GIESE, A. C.: Cell physiology (Saunders, Philadelphia 1968).

GLASSTONE, S.: Introduction to electrochemistry (Van Nostrand, New York 1960).

GOLDSCHMIDT, V. M.: Geochemical distribution of the elements. Skriften Norske Videnskaps Akad. (Oslo). J. Mat. Natur. Kl. *8* (1926).

HARDY, W. B.: On the structure of cell protoplasm. J. Physiol., Lond. *24:* 158–210 (1899).

HARDY, W. B.: Colloidal solutions. The globulins. J. Physiol., Lond. *33:* 251–335 (1905).

HAZLEWOOD, C. F. and NICHOLS, B. L.: Changes in muscle sodium, potassium, chloride, water and voltage during maturation in the rat; an experimental and theoretical study. Johns Hopk. med. J. *125:* 119–133 (1969).

HAZLEWOOD, C. F.; NICHOLS, B. L., and CHAMBERLAIN, N. F.: Evidence of a minimum of two phases of ordered water in skeletal muscle. Nature, Lond. *222:* 747–758 (1969).

HEIJENVOORT, J. VAN: From Frege to Gödel (Harvard University Press, Cambridge, Mass. 1967).

HELMHOLTZ, H. VON: Über die Erhaltung der Kraft (Berlin 1847).

HENDERSON, L. J.: Blood; a study in general physiology (Yale, New Haven 1928).

HILL, A. V.: Living machinery (Bell, London 1944).

HILL, A. V.: The mechanics of voluntary muscle. Lancet *ii:* 947–951 (1951).

HOFMEISTER, F.: Zur Lehre von der Wirkung der Salze. Arch. exp. Path. Pharmakol. *24:* 247–260 (1888).

HOYLE, G. and SMYTH, T.: Neuromuscular physiology of giant muscle fibers of a barnacle, *Balenus Nubilus* Darwin. Comp. Biochem. *10:* 291–314 (1933).

HUXLEY, T. H.: The physical basis of life (lay sermon, 1868); quoted in HARDY, Collected papers, (Cambridge University Press, London 1936).

JOSEPH, N. R.: Efficiency of ion transport mechanisms. Nature, Lond. *209:* 398–399 (1966a).

JOSEPH, N. R.: Intracellular hydration of ions in relation to age. Gerontologia, Basel *12:* 155–173 (1966b).

JOSEPH, N. R.: Electrolyte balance and cellular aging. Gerontologia, Basel *14:* 142–159 (1968).

JOSEPH, N. R.: Physical chemistry of aging (Karger, Basel 1971a].

JOSEPH, N. R.: Dependence of electrolyte balance on growth and development of cells and tissues; in ELDEN A treatise on skin (Wiley, Interscience, New York 1971b).

JOSEPH, N. R.; ENGEL, M. B., and CATCHPOLE, H. R.: Distribution of sodium and potassium in certain cells and tissues. Nature, Lond. *191:* 1175–1178 (1961).

JOSEPH, N. R.; ENGEL, M. B., and CATCHPOLE, H. R.: Chemical potentials and electrical potentials in biological systems. Nature, Lond. *203:* 931–933 (1964).

JOSEPH, N. R.; ENGEL, M. B., and CATCHPOLE, H. R.: Distribution of ions in biological systems. Nature, Lond. *206:* 6–11 (1965).

JOULE, J. P.: Scientific papers (London 1884–1887).

KATZ, J.: Die mineralischen Bestandteile des Muskelfleisches. Pflugers Arch. ges. Physiol. *63:* 1–85 (1896).

KAVENAU, J. L.: Water and solute-water interactions (Holden-Day, San Francisco 1964).

KLOTZ, I.: Role of water structure in macromolecules. Fed. Proc. Suppl. *15:* 324–333 (1965).

KOSSEL, A.: The protamines and histones (1893) (Longmans-Green, New York 1928).

LAIDLER, K. J. and PEGIS, C.: The influence of dielectric saturation on the thermodynamic properties of aqueous ions. Proc. roy. Soc. Lond. *A241:* 80–92 (1957).

LEWIS, G. N. and RANDALL, M.: Thermodynamics (McGraw-Hill, New York 1961).

LOEB, J,: The dynamics of living matter (Columbia University Press, New York 1906).

LOEB, J.: Proteins and the theory of colloidal behavior (McGraw-Hill, New York 1924).

LORENZ, K.: Studies in animal and human behavior, vol. 1 and 2 (Harvard University Press, Cambridge, Mass. 1970 and 1971).

LOTKA, A. J.: Elements of physical biology (Williams & Wilkins, Baltimore 1924).

LOVE, R. M.: The chemical biology of fishes (Academic Press, New York 1970).

MACALLUM, A. B.: The inorganic composition of the blood in vertebrates and invertebrates and its origin. Proc. roy. Soc. Lond. *B82:* 602 – 624 (1910).

MACALLUM, A. B.: The paleochemistry of the body fluids and tissues. Physiol. Rev. *6:* 216–257 (1926).

MARSHALL, E. K.: The comparative physiology of the kidney in relation to theories of renal secretion. Physiol. Rev. *14:* 133–159 (1934).

MAYER, J. R.: Bemerkungen über die Kräfte der unbelebten Natur. Justus Liebigs Ann. Chem. *42:* 233 (1842).

MEYER, K. H.: Thermoelastic properties of several biological systems. Proc. roy. Soc. *B139:* 498–505 (1951–1952).

MIESCHER, F.: Das Protamin, eine neue organische Base aus den Samenfäden des Rheinlachses. hses. Ber. dtsch. chem. Ges. *7:* 376–387 (1874).

MORROW, J. E.: in GRAY Encyclopedia of the biological Sciences (Reinhold, New York 1961).

NASTUK, W. L.: The electrical activity of the muscle cell membrane at the neuromuscular membrane. J. cell. comp. Physiol. *42:* 249–272 (1953).

NASTUK, W. L. and HODGKIN, A. L.: The electrical activity of single muscle fibers. J. cell. Comp. Physiol. *35:* 39–73 (1950).

NICHOLLS, J. G.: The electrical potential of denervated skeletal muscle. J. Physiol., Lond. *131:* 1–12 (1956).

OPARIN, A. I.: Origin of life (Macmillan, New York 1938).

PARRY, G.: Osmotic and ionic changes in blood and muscle in migrating salmonids. J. exp. Biol. *38:* 311–427 (1961).

PAULING, L.: The nature of the chemical bond (Cornell University Press, Ithaca 1944).

PEIRCE, C. S.: Man's glassy essence (1892); reprinted in Chance, love and logic (Harcourt, Brace & World, New York 1923).

POTTS, W. T. W.: The inorganic composition of the blood of *Mytilus edulis* and *Anodonta cygnea.* J. exp. Biol. *31:* 376–385 (1955).

POTTS, W. T. W.: The inorganic and amino acid composition of some lamellibranch muscles. J. exp. Biol. *35:* 749–764 (1958).

RICHARDS, A. N.: Processes of urine formation. Proc. roy. Soc. Lond. *B126:* 398–432 (1938).

RINGER, S.: Concerning the influence exerted by each of the constituents of the blood on the contraction of the ventricle. J. physiol., Lond. *3:* 380–393 (1880–1882).

ROBERTSON, J. D.: Studies on the chemical composition of muscle tissue. I. The muscles of the hagfish *Myxina glutinosa* L. and the Roman eel, *Maecena helena* L. J. exp. Biol. *37:* 879–888 (1960).

ROBERTSON, J. D.: Studies on the chemical composition of muscle fibre. II. The abdominal flexor muscle of the lobster *Nephrops Norvegicus.* J. exp. Biol. *38:* 707–728 (1961).

ROPES, M. W. and BAUER, W.: Synovial fluid changes in joint disease (Harvard University Press, Cambridge, Mass. 1953).

RUBNER, M.: Calorimetrische Untersuchungen. Z. Biol. *1894:* 73–142.

SARTON, G.: A history of science (Harvard University Press, Cambridge, Mass. 1952).

SCHRODINGER, E.: What is life (Cambridge University Press, Cambridge 1944).

SHAW, J.: Ionic regulation of the muscle fibres of *Carcinus maenas.* I. The electrolyte composition of single fibres. J. exp. Biol. *32:* 385–396 (1955).

SHAW, J.: Further studies on ionic regulation in the muscle fibres of *Carcinus maenas.* J. exp. Biol. *35:* 902–910 (1958a].

SHAW, J.: Osmoregulation in the muscle fibres of *Carcinus maenas.* J. exp. Biol. *35:* 920–929 (1958b).

SHOCK, N. W. and YIENGST, M. J.: Age changes in basal respiratory measurements and metabolism in males. J. Geront. *10:* 31–40 (1955).

SORENSEN, S. P. L.: Studies on proteins. C. R. Lab. Carlsberg *12:* 1–67 (1915).

SORENSEN, S. P. L.: Die Konstitution der löslichen Proteinstoffe als reversibel dissoziables Komponentensystem. Kolloid Z. *53:* 101–128 (1930).

STENSEN, N.: Observationes anatomiques (1662). Lectures on the history of physiology (Dover, New York 1971).

SVEDBERG, T.: Colloid chemistry (Chemical Catalog, New York 1928).

TANNER, J. M.: The physique of the Olympic athlete (Allen & Unwin, London 1964).

TRAUTWEIN, W.; ZINK, K., und KAYSER, K.: Über Membran- und Aktionspotentiale einzelner Fasern des Warmbluterskeletmuskels und ihre Veränderung bei der Ischämie. Pflügers Arch. ges. Physiol. *257:* 20–34 (1953).

WHEELER, W. M.: Social life among the insects (Harcourt, Brace & World, New York 1923).

WHITEHEAD, A. N.: Science and the modern world (Macmillan, New York 1925).

WHITEHEAD, A. N.: Process and reality (Macmillan, New York 1929).

WIDDOWSON, E. M. and DICKERSON, J. W. T.: The effect of growth and function on the chemical composition of soft tissues. Biochem. J. *74:* 247–257 (1960).

WIDDOWSON, E. M. and DICKERSON, J. W. T.: Chemical composition of the body; in COMAR and BRONNER Mineral metabolism, vol. 2A (Academic Press, New York 1964).

Subject Index